Get Wise!
Mastering
Math Word
Problems

Get Wise!™

MASTERING
Math Word
Problems

Maureen Steddin

THOMSON

PETERSON'S

Australia • Canada • Mexico • Singapore • Spain • United Kingdom • United States

THOMSON
PETERSON'S

About The Thomson Corporation and Peterson's

The Thomson Corporation, with 2002 revenues of US$7.8 billion, is a global leader in providing integrated information solutions to business and professional customers. The Corporation's common shares are listed on the Toronto and New York stock exchanges (TSX: TOC; NYSE: TOC). Its learning businesses and brands serve the needs of individuals, learning institutions, corporations, and government agencies with products and services for both traditional and distributed learning. Peterson's (www.petersons.com) is a leading provider of education information and advice, with books and online resources focusing on education search, test preparation, and financial aid. Its Web site offers searchable databases and interactive tools for contacting educational institutions, online practice tests and instruction, and planning tools for securing financial aid. Peterson's serves 110 million education consumers annually.

For more information, contact Peterson's, 2000 Lenox Drive, Lawrenceville, NJ 08648; 800-338-3282; or find us on the World Wide Web at www.petersons.com/about.

Production Editor: Teresina Jonkoski; Page Designer: Linda M. Williams; Illustrations: Gary Van Dzura and Chris Allan; Production Director: Bernadette Webster; Editorial Director: Del Franz

ISBN: 0-7689-1600-3

Printed in Canada

10 9 8 7 6 5 4 3 2 1   06 05 04

First Edition

Man, this page is boring!

Contents

introduction

Let me get this straight: you guys wrote a whole book about *word problems*?

Yes, this *entire* book is dedicated to word problems! And, if the thought of pages and pages of questions that combine math and reading (the worst of both worlds) makes your head spin, you've come to the right place. We're going to make word problems seem easy—so no more head-spinning, nightmares, or palpitations.

You may not realize it, but you've got to deal with a lot of word problems while you are in school. Not only do you see them on your classroom tests, but they'll pop up on state standardized tests and college entrance exams like the SAT and ACT Assessment. What you will soon learn from this book, though, is that word problems are just real-life situations involving math. Believe it or not, word problems are actually relevant to your life outside the classroom.

Not convinced? OK. Here's an example. Calculating what 85% of 60 is sounds pretty academic, right? But, if you find out that the $60 sweater you've been dying to get has been discounted by 15%, suddenly figuring out the final sales price (and whether you have enough money to buy it) seems pretty important.

Here's another one. Determining the fifth number that when averaged with 84, 78, 81, and 90 will produce an average of 85 may seem pretty meaningless. But if you received grades of 84, 78, 81, and 90 on your first four math tests and want to figure out what grade you need on your fifth test to make your average 85 (not to mention to make your teacher and parents happy), this calculation seems more worthwhile.

Don't think that every single math word problem you see will have direct personal meaning for you, though. Some questions are less relevant to your everyday experience than others, but the skills that they build are the same. They'll show you how the math concepts and formulas that you learn in the classroom have use in the real world and will generally help you to become a better problem-solver.

Now, let's focus on this book. There are thirteen chapters, each covering a different concept. You don't need to work through them in order, and you can jump to the type of problems that you need the most help with. However, we believe in the expression "practice makes perfect" and recommend that you check out every chapter so you can test your skills. No matter how you decide to use this book, you should start with Chapter One since it's the basis for the rest of the problems in the book.

A number of years ago, there was a talking version of a certain poseable fashion doll (who shall remain nameless but whose favorite color is pink) that said, "Math is hard," as if it hurt her head just to think about adding 2 and 2.

Oh please, I hate that helpless girl stuff. My hate-hate relationship with math is not because it's hard—it's because it's boring and irrelevant to my life!

OK, Chi. We can't promise that word problems will be as much fun as a trip to the mall or an afternoon hanging out with your friends. We can, however, try to show you how math *does* apply to your life and can even be useful.

Enough of this intro stuff. It's time to *Get Wise!*

Petersons.com/publishing

Check our Web site at www.petersons.com/ publishing to see if there is any new information regarding any revisions or corrections to the content of this book. We've made sure the information in this book is accurate and up-to-ate; however, the content may have changed since the time of publication.

chapter 1

It's All in the Translation

UNDERSTANDING TRANSLATION

Translation? Did I miss something here? I thought this was about math, not Spanish or French.

While you are probably thinking the same thing as Chi, translation is a big part of dealing with math problems. Instead of helping you find the bathroom in an international city, translating in math will help you put a word problem into mathematical terms so you can find its solution.

The chart below gives you some of the terms that come up a lot in word problems. Consider it your pocket dictionary between English and math. You'll probably recognize a number of the translations, but it's still a good reference tool to have. As you can see, there are multiple terms in English that translate to the same mathematical concept or operation, so make sure that you become familiar with all of them.

English	Math	Example	Translation
What, a number	$x, n,$ etc.	Three more than a number is 8.	$n + 3 = 8$
Equals, is, was, has, costs	=	Danny **is** 16 years old. A CD **costs** 14 dollars.	$d = 16$ $c = 14$
Is greater than Is less than At least, minimum At most, maximum	> < ≥ ≤	Jenny **has more** money than Ben. Ashley's age **is less than** Nick's. There are **at least** 30 questions on the test. Chi can invite **a maximum** of 15 people to her party.	$j > b$ $a < n$ $t \geq 30$ $c \leq 15$
More, more than, greater than, added to, total, sum, increased by, together	+	Kecia has 2 **more** video games than John. Kecia and John have a **total** of 11 video games.	$k = j + 2$ $k + j = 11$
Less than, smaller than, decreased by, difference, fewer	−	Tyler has 3 **fewer** CDs than Carson. The **difference between** Jenny's and Ben's savings is $75.	$t = c - 3$ $j - b = 75$
Of, times, product of, twice, double, triple, half of, quarter of	×	Emma has **twice** as many books as Justin. Justin has **half** as many books as Emma.	$e = 2 \times j$ or $e = 2j$ $j = e \times \dfrac{1}{2}$ or $j = \dfrac{e}{2}$

English	Math	Example	Translation
Divided by, per, for, out of, ratio of ___ to ___	/, ÷	Sophia has $1 **for** every $2 Lizzy has.	$s = l \div 2$ or $s = \dfrac{l}{2}$
		The **ratio of** Lizzy's savings **to** Sophia's savings is 2 to 1.	$\dfrac{l}{s} = \dfrac{2}{1}$

When you study a foreign language, there's the chance of a trip abroad, or at least of getting to eat exotic foods in class. What do we get for translating English to math?

Translating may seem a bit intimidating, but it's really not that hard. First, you assign variables to the unknowns in the question. It usually makes sense to use the first letter of the unknown as the variable (this is where the chart comes in—check it out for examples). Don't use the same variable to represent more than one unknown, though. After you have assigned variables, translate the information you're given piece by piece. Make sure to think about the question logically as you translate.

There's a classic translation error that people make all the time. We're telling you about it now to spare you from falling into this trap yourself. Look at the following English statement and think about how to translate it into math:

Example: Translate the following statement: "Jennifer has 10 fewer DVDs than Brad."

Classic Translation Mistake: $j - 10 = b$

Do you know why this is wrong? Give up? Well, you can't just translate from left to right. You have to make sure that the symbols for mathematical operations (like +, −, ×, and ÷) are acting on the correct variable. The statement was translated incorrectly to:

Original	Translation
Jennifer	j
has 10 fewer DVDs	$- 10$
than Brad.	$= b$
	$j - 10 = b$

This translation is wrong because "has" is translated as a minus sign instead of an equal sign, and "than" is incorrectly translated as an equal sign.

Correct Translation

$$j = b - 10$$

Think about the statement logically to help you come up with this translation. Even though the words "10 fewer DVDs" come after Jennifer, they are referring to Brad, which is what the "than" tells you. Here's a breakdown of the correct translation:

Original	Translation
Jennifer has	$j =$
10 fewer DVDs than Brad.	$b - 10$
	$j = b - 10$

So now I'm expected to combine translation skills, math knowledge, and logical thinking. Anything else you want to pile on top of that?

Get Wise!

Try the following problems involving translation.

1. Translate the following statement mathematically: "A DVD costs $10 more than a CD." Let d be the price of a DVD and c be the price of a CD.

   (A) $d + 10 = c$ (D) $10d = c$

   (B) $d = c + 10$ (E) $d = \dfrac{c}{10}$

   (C) $d + c = 10$

2. Which of the following algebraic expressions best translates the statement: "8 decreased by half a number is equal to 5"?

   (A) $8 - n = 5$ (D) $8 - \dfrac{n}{2} = 5$

   (B) $8 - 2n = 5$ (E) $8 - \dfrac{1}{2}(5) = n$

   (C) $8 - \dfrac{1}{2} = 5$

3. Translate the following statement mathematically: "Jan Brady had 3 fewer dates than Marcia Brady."

   (A) $j = 3 - m$ (D) $3 - j = m$
   (B) $j - 3 = m$ (E) $j - m = 3$
   (C) $j = m - 3$

Poor Jan. It's always Marcia, Marcia, Marcia . . .

4. Which of the following algebraic expressions best translates the statement: "Together Cain and Abel have at least \$35"?

(A) $c + a = 35$ (D) $c + a > 35$

(B) $c + a < 35$ (E) $c + a \geq 35$

(C) $c + a \leq 35$

5. Translate the following statement mathematically: "If Maureen's hourly wage were increased by \$6, she'd be earning twice what she is now."

(A) $m + 6 = 2m$ (D) $m - 6 = 2m$

(B) $6m = m + 2$ (E) $m + 6 = m + 2$

(C) $m + 6 = \dfrac{m}{2}$

How Wise?

Check your answers on page 25.

"IN TERMS OF"

So far, you've used a different variable to represent each unknown. Sometimes, though, you can represent different quantities in terms of one variable. This doesn't mean that you just use the same letter for each unknown. It only works if you're given information about the relative sizes of the unknowns. If you're told, for example, that Henry has three times as much money as Cole, you could represent Cole's money as c and Henry's as $3c$.

If Henry had three times as much money as me, he'd still be in bad shape. Three times nothing is still nothing. See, some of this math information is seeping into my brain.

Example: Clay got 10 fewer votes than Kimberly. Reuben got three times as many votes as Clay. The three contestants received a total of 90 votes. Write an equation in one variable that can be used to solve for the number of votes Kimberly received.

Solution: Pick which unknown will be represented by the variable. In this case you want an equation that you can solve for the number of votes Kimberly received. This means that when the equation is solved, the value of the variable must be the number of votes she received. So let k be the number of votes Kimberly received.

Now you need to represent the other two unknowns in terms of k. You're told that Clay got 10 fewer votes than Kimberly. Since Kimberly has k votes, Clay has 10 fewer than that, or $k - 10$.

You're told that Reuben got three times as many votes as Clay. Since Clay has $k - 10$ votes, Reuben has three times that, or $3(k - 10)$.

Next, write an equation with all these expressions. You're told that the total number of votes that the contestants received is 90. This means that the sum of the three variable expressions will be 90:

$$k + (k - 10) + 3(k - 10) = 90$$

Get Wise!

Try the following problems involving translating "in terms of."

1. Bobby ate 3 more slices of pizza than Peter. Greg ate half as many slices of pizza as Bobby. The boys ate a total of 12 slices of pizza. Which of the following equations could be used to solve for the number of slices of pizza that Peter ate?

   (A) $3p = 12$

   (B) $p + (p + 3) + 2p = 12$

   (C) $p + (p + 3) + \dfrac{2}{p} = 12$

   (D) $p + (p + 3) + \dfrac{p + 3}{2} = 12$

   (E) $p + (3 - p) + \dfrac{p}{2} = 12$

Sounds like Bobby really likes pizza!

2. Eileen and Julie went shopping. Eileen spent $10 less than twice as much as Julie. If the two girls spent a total of $140, which of the following equations could be solved for the amount that Julie spent?

   (A) $x + 2(x - 10) = 140$

   (B) $x + (2x - 10) = 140$

   (C) $x + (x - 10) = 140$

   (D) $(x - 10) + (2x - 10) = 140$

   (E) $2x - 10 = 140$

3. A bag contains a total of 27 lollipops, all of which are either lemon, grape, or cherry. Twice as many are lemon as grape. Three fewer lollipops than lemon are cherry. Which equation can be used to solve for the number of grape lollipops in the bag?

    (A) $n + 2n + (2n - 3) = 27$

    (B) $n + 2n + (n - 3) = 27$

    (C) $n + 2n + (3 - 2n) = 27$

    (D) $n + 3n + (3n - 2) = 27$

    (E) $n + 2n + 3n = 27$

4. Reese is 8 years more than half as old as Rob. Their total ages sum to 50. Which of the following equations can be used to solve for Rob's age?

    (A) $R + \left(\dfrac{R}{2} - 8 \right) = 50$

    (B) $R + \left(\dfrac{R}{2} + 8 \right) = 50$

    (C) $R + \dfrac{R + 8}{2} = 50$

    (D) $\dfrac{R}{2} + \dfrac{R}{6} = 50$

    (E) $R + \left(\dfrac{2}{R} + 8 \right) = 50$

5. Ryan has 5 fewer than three times as many games for Nintendo™ than he has for PlayStation™. If Ryan has a total of 23 games, which of the following equations can be used to solve for the number of PlayStation games he has?

(A) $(g - 5) + 3g = 23$

(B) $g + (5 - 3g) = 23$

(C) $g + (3g - 5) = 23$

(D) $5g + 3g = 23$

(E) $g + (5g - 3) = 23$

How Wise?

Check your answers on page 25.

EQUAL RIGHTS

Translating inequalities is just like translating equations. The only difference is that you use an inequality sign instead of an equal sign. But since there are four different inequality signs (<, >, ≤, and ≥), you might get a bit confused as to which to use where.

I know a lot about inequalities. My parents give me way more chores than they give my brother. I am not even allowed to stay out as late as my friends. And we kids don't get to vote either.

Example: Emma is a salesperson who is paid $220 a week, plus a commission of $15 for each cell phone she sells. If *c* represents the number of cell phones, write an inequality that can be used to determine how many cell phones Emma must sell in order to earn at least $400 a week.

Solution: As always, read through the question carefully to get a sense of the situation. This question involves Emma's weekly salary, which changes depending upon how many cell phones she sells. You're asked to write an inequality that can be used to figure out the number of phones she must sell each week to earn at least $400.

The first step of a translation is usually to assign variables to any unknowns. In this question there's just one unknown, the number of cell phones sold. You're already told to use *c* for this value, so that much is done for you.

Next, decide which inequality sign you will use. If you have the super power of a photographic memory, you'll have total recall of the translation chart from the start of this lesson. Then you'd know that "at least" translates to ≥.

If you're a mere mortal like the rest of us, you can use some logic to help you figure out which sign to use. Emma wants to earn at least $400 a week. This means that she'd be happy earning an amount equal to $400 or any amount greater than $400. This means that the correct inequality should contain ≥ 400.

Now translate the rest of the question. Emma earns a set fee of $220 each week, and $15 for each cell phone she sells. Using c as the number of cell phones sold, this translates to $220 + 15c$.

So the entire inequality is $220 + 15c \geq 400$.

Emma should have no problem selling enough cell phones. Since they are making cooler cell phones all the time, everyone I know wants a new one.

Get Wise!

Try the following problems where you'll have to translate inequalities.

1. Jen has read 8 pages of her social studies chapter. She wants to read a minimum of y pages before she goes to bed. If z represents the additional pages she must read, which of the following best represents the number of pages Jen must read before going to bed?

(A) $8 + z > y$ (D) $8 + z \leq y$

(B) $8 + z < y$ (E) $8 + z = y$

(C) $8 + z \geq y$

2. A school is having a special event to honor successful alumni. The event will cost $500, plus an additional fee of $85 for each alum who is honored. If the school can afford to spend no more than $1,000 total on the event, which of the following best represents the number of alumni they can honor?

(A) $85a \geq 1,000$ (D) $500 + 85a \leq 1,000$

(B) $85a \leq 1,000$ (E) $500a + 85 < 1,000$

(C) $500 + 85a \geq 1,000$

Someday when I'm rich and famous I'll go back to my school to visit—that is, if I'm not too busy jet-setting around.

3. Karen has p dollars and Jack has q dollars. Together, they still have less than they need to buy a camera costing r dollars. Which of the following best expresses this situation?

(A) $p + q > r$ (D) $p + q \leq r$

(B) $p + q < r$ (E) $p + q = r$

(C) $p + q \geq r$

4. Bob the beagle is allowed to have a maximum of m treats per day. He has already had b treats today. If t is the number of treats Bob is still allowed to have, which of the following best represents this situation?

(A) $b - t \leq m$ (D) $b + t \geq m$

(B) $b - t \geq m$ (E) $bt = m$

(C) $b + t \leq m$

5. Archie collected a gum wrappers and gave j away to Jughead. Archie still has more than the x gum wrappers he needs to send away for a pair of X-ray vision glasses. Which of the following best represents this situation?

(A) $a - j > x$ (D) $a + j \geq x$

(B) $a + j > x$ (E) $a - j = x$

(C) $a - j \geq x$

Those X-ray vision glasses are a load of garbage. You have to send in a million wrappers, and then wait forever for them to arrive, and when you finally get them they don't work at all. That's what I heard anyway—it's not like I'd get something like that myself.

How Wise?

Check your answers on page 27.

THE PRICE ISN'T ALWAYS RIGHT

Some questions include a situation where there is more than one cost. One of them is fixed, meaning that you pay a specific amount once. The other is variable, meaning that how much you pay depends on how many miles you drive, or how many minutes you talk, and so on. The questions don't always have to be about the price of something, though. The inequality example we looked at earlier fits into this category. Emma earned a fixed fee for working each week, and then a variable amount that depended upon how many phones she sold. A common mistake in these problems is mixing up the fixed and variable costs. We wouldn't want you to go and do something silly like that, so we'll show you how to handle one of these questions.

Example: For a certain billing plan, a cell phone company charges $25 each month and $0.15 for each minute used over the plan limit. Write an equation that expresses the relationship between the total monthly charges, c, and minutes used over the plan limit, m.

Solution: First, identify what each variable stands for. The question is about a monthly cell phone bill. The total amount of the bill is represented by c, and the number of minutes used over the plan limit is represented by m.

Now figure out which value in the question is the fixed cost and which is linked to the variable cost. Since there is a one-time charge of $25, this is the fixed cost. Since there is a charge of $0.15 for each minute used over the plan limit, this is the variable cost. The variable cost is the number of minutes used over the plan limit, multiplied by the price per minute. The number of minutes is m and the price per minute is $0.15. So the variable cost of the call is $.15m$.

The total amount of the phone bill c is the result of adding together the fixed and variable costs:

$$c = 25 + .15m$$

Get Wise!

Try the following problems with multiple costs.

1. A Web site charges a $5 handling fee for each order, plus a shipping fee of $0.55 for each pound that the order weighs. Which equation best expresses the relationship between the number of pounds of the order shipped, p, and the total shipping and handling fees, s?

(A) $s = 5 + p$

(B) $s = 5p + .55p$

(C) $s = 5 + .55 + p$

(D) $s = 5p + .55$

(E) $s = 5 + .55p$

2. A large pizza costs $12 plus $0.75 for each topping. Which of the following equations represents the relationship between the price of a large pie, p, and the number of toppings it contains, t?

(A) $p = 12 + .75$

(B) $p = 12 + .75t$

(C) $p = 12t + .75$

(D) $p = 12t + .75t$

(E) $p = .75(12 + t)$

3. A computer repair company charges $50 for a service call plus $25 for each hour of work. Which of the following equations represents the relationship between the bill, b, for a service call, and the number of hours spent on the call, h?

(A) $25(50 + h) = b$

(B) $50h + 25 = b$

(C) $50 + 25h = b$

(D) $50h + 25h = b$

(E) $50 + 25 = b$

4. A taxicab charges $2.00 for a ride, plus $0.40 for each mile, *m*, driven. The cab driver hopes to make a minimum of $5.00 from his next customer. Which of the following could be used to solve for the value of *m*?

(A) $2m + .40m = 5$

(B) $2 + .40m \geq 5$

(C) $2 + .40m \leq 5$

(D) $2m + .40 \geq 5$

(E) $2m + .40 \leq 5$

I guess the cab driver will be taking the scenic route with his next customer.

5. The Griswolds want to rent a car for a family vacation. The rental costs $100 per week plus a charge of $0.15 per each mile, *m*, driven. The family can spend a maximum of $200 on the car rental. Which of the following best represents this situation?

(A) $100m + .15m \geq 200$

(B) $100 + .15m \geq 200$

(C) $100m + .15 \geq 200$

(D) $100 + .15m \leq 200$

(E) $100m + .15 \leq 200$

How Wise?

Check your answers on page 28.

WISE NOTE

This chapter focused on translating English into algebra. That's why the problems in it required you to translate into mathematical expressions, equations, or inequalities, but not to solve for any specific values. You will come across many word problems where all that you have to do is translate. In some cases, however, you will need to solve an equation or inequality.

Cool! Sounds like we really caught a break here. We could have had to translate AND solve all of these questions.

To solve an equation, isolate the variable on one side of the equation, by "undoing" everything that has been done to it.

To solve an inequality, follow the method for solving an equation. The only difference is that if you multiply or divide by a negative number, you must flip the inequality sign.

A Word to the Wise

- ★ Translating English to math lets you put a problem into mathematical terms so that you can find its solution.

- ★ Assign variables and then translate a statement piece by piece. Make sure that your translation follows the logic of the situation.

- ★ If you're given information about the relative sizes of the unknowns, you may be able to express them all in terms of the same variable.

- ★ If you're having trouble figuring out which inequality sign to use, try to let the logic of the situation guide you.

- ★ When translating a question that includes multiple costs, be careful not to mix up the fixed and variable costs.

ANSWERS TO CHAPTER 1: PRACTICE EXERCISES

Understanding Translation (Page 9)

1. **The correct answer is (B).** The problem tells you that d is the price of a DVD and c the price of a CD. "A DVD costs" translates to $d =$. "$10 more than a CD" translates to $c + 10$. This makes the entire equation $d = c + 10$.

2. **The correct answer is (D).** Let n be the number. "8 decreased by" translates to $8 -$. "Half a number" translates to $\frac{n}{2}$. "Is equal to 5" translates to $= 5$. This makes the entire equation $8 - \frac{n}{2} = 5$.

3. **The correct answer is (C).** Let j be the number of dates Jan had and m be the number of dates Marcia had. "Jan had" translates to $j =$. "3 fewer dates than Marcia" translates to $m - 3$. This makes the entire equation $j = m - 3$.

4. **The correct answer is (E).** Let c be the amount of money that Cain has and a be the amount of money that Abel has. "Together Cain and Abel" translates to $c + a$. "Have at least $35" translates to ≥ 35. This makes the entire inequality $c + a \geq 35$.

5. **The correct answer is (A).** Let m be Maureen's hourly wage. "Maureen's hourly wage increased by 6" translates to $m + 6$. "She'd be earning twice what she does now" translates to $= 2m$. This makes the entire equation $m + 6 = 2m$.

Translating "In Terms Of" (Page 13)

1. **The correct answer is (D).** You're asked for an equation that can be solved for the number of slices of pizza that Peter ate, so use a variable to represent Peter's share. All of the answer choices use the variable p, so let p be the number of slices that Peter ate. Bobby ate 3 more pieces than Peter, so he ate $p + 3$. Greg ate half as many slices

as Bobby, so he ate $\dfrac{p+3}{2}$. They ate a total of 12 slices, so sum these

three variable expressions to 12: $p+(p+3)+\dfrac{p+3}{2}=12$.

2. **The correct answer is (B).** You're asked for an equation that can be solved for the amount of money that Julie spent, so use a variable to represent Julie's share. All of the answer choices use the variable x, so let x be the amount of money that Julie spent. Eileen spent \$10 less than twice as much as Julie; twice as much as Julie is $2x$, so \$10 less than that is $2x - 10$. They spent a total of \$140, so sum these two variable expressions to 140: $x + (2x - 10) = 140$.

3. **The correct answer is (A).** You're asked for an equation that can be solved for the number of grape lollipops in the bag, so use a variable to represent the grape pops. All of the answer choices use the variable n, so let n be the number of grape lollipops in the bag. Twice as many pops are lemon as grape, so there are $2n$ lemon pops. Three fewer pops than lemon are cherry, so there are $2n - 3$ cherry pops. There are a total of 27 pops in the bag, so sum these three variable expressions to 27: $n + 2n + (2n - 3) = 27$.

4. **The correct answer is (B).** You're asked for an equation that can be solved for Rob's age, so use a variable to represent his age. All of the answer choices use the variable R, so let R be Rob's age. Reese is 8

years more than half as old as Rob; half as old as Rob is $\dfrac{R}{2}$, so 8

years more than this is $\dfrac{R}{2}+8$. Their ages total 50, so sum these two

variable expressions to 50: $R+\left(\dfrac{R}{2}+8\right)=50$.

5. **The correct answer is (C).** You're asked for an equation that can be solved for the number of PlayStation games that Ryan has. All of the answer choices use the variable g, so let g be the number of PlayStation games. Ryan has 5 fewer than three times as many Nintendo games as PlayStation games; three times as many is $3g$, so 5 fewer than this is $3g - 5$. Ryan has a total of 23 games, so sum these two variable expressions to 27: $g + (3g - 5) = 23$.

Translating Inequalities (Page 18)

1. **The correct answer is (C).** Jen wants to read a minimum of y pages before she goes to bed. That means that the number of pages she reads must be equal to or greater than y. So the correct inequality should contain $\geq y$. Jen has read 8 pages and needs to read z more; this translates to $8 + z$. This makes the entire inequality $8 + z \geq y$.

2. **The correct answer is (D).** The school can spend no more than $1,000. This means that the amount of money that it can spend is equal to or less than $1,000. So the correct inequality should contain $\leq 1,000$. The event will include a fixed cost of $500 and a variable cost of $85 per alum honored. Since all the equations use the variable a, let a be the number of alumni honored at the event. Then you can translate the total cost of the event to $500 + 85a$. This makes the entire inequality $500 + 85a \leq 1,000$.

3. **The correct answer is (B).** The total amount of money that Karen and Jack have is less than they need to buy the camera for r dollars. So the correct inequality should contain $< r$. Since Karen has p dollars and Jack has q dollars, together they have a total of $p + q$. This makes the entire inequality $p + q < r$.

4. **The correct answer is (C).** Bob the beagle can have a maximum of m treats in a day. This means that the total number of treats he can have must be equal to or less than m. So the correct inequality should contain $\leq m$. Bob, who has already eaten b treats, is allowed to eat t more; this translates to $b + t$. This makes the entire inequality $b + t \leq m$.

5. **The correct answer is (A).** The number of wrappers that Archie has left after giving some away is more than the x he needs for his offer. This means that the correct inequality should contain $> x$. Since Archie has a wrappers and gives j away, he has a total of $a - j$ wrappers left. This makes the entire inequality $a - j > x$.

Problems with Multiple Costs (Page 21)

1. **The correct answer is (E).** The total shipping and handling fees, s, is the sum of the fixed and variable costs. The fixed cost is $5. The variable cost is $0.55 for each of p pounds. This makes the variable cost $.55p$. So the total equation is $s = 5 + .55p$.

2. **The correct answer is (B).** The total cost of a large pizza, p, is the sum of the fixed and variable costs. The fixed cost is $12. The variable cost is $0.75 for each of t toppings. This makes the variable cost $.75t$. So the total equation is $p = 12 + .75t$.

3. **The correct answer is (C).** The total bill b for a service call is the sum of the fixed and variable costs. The fixed cost is $50. The variable cost is $25 for each of h hours. This makes the variable cost $25h$. So the total equation is $50 + 25h = b$.

4. **The correct answer is (B).** The cab driver wants to make a minimum of $5 on his next fare. This means that he will make an amount equal to or greater than $5. So the correct inequality should contain ≥ 5. The cost of the cab ride is the sum of the fixed and variable costs. The fixed cost is $2. The variable cost is $0.40 for each of m miles driven. This makes the variable cost $.40m$. So the total inequality is $2 + .40m \geq 5$.

5. **The correct answer is (D).** The Griswolds want to spend a maximum of $200 on the car rental. This means that the amount that they spend on the car rental will be equal to or less than $200. So the correct inequality should contain ≤ 200. The cost of the car rental is the sum of the fixed and variable costs. The fixed cost is $100. The variable cost is $0.15 for each of m miles driven. This makes the variable cost $.15m$. So the total inequality is $100 + .15m \leq 200$.

chapter 2

Is There a Problem with My Number?

UNDERSTANDING PROBLEMS ABOUT NUMBERS

Lots of word problems describe real-life situations. The ones in this chapter, though, are all about numbers. Not very exciting perhaps, or relevant to your daily life, but you need to be prepared for them since they turn up on lots of tests like the SAT and ACT Assessment. The translating and equation-solving skills involved apply to all word problems, so practicing them will be very useful (practice makes perfect . . .).

So it'll be good for me even though I don't enjoy it. Like taking medicine or eating Brussels sprouts.

You'll come across a lot of math terms when you translate word problems about numbers. You'll need to be familiar with what they mean in order to set up and solve them correctly. You'll probably be familiar with a lot of them already; many were in the chart in Chapter One.

Example: The product of a number and 5 is equal to the sum of that number and 12. What is the number?

Solution: Translate the statement to a math equation, and then solve the equation. If you have time, it's a good idea to check your work.

Translate: If you don't know what the words "product" and "sum" mean, you'll be scratching your head on this one. It's important to know the terms that represent the four basic arithmetic operations:

Sum is the result of **addition**.

Difference is the result of **subtraction**.

Product is the result of **multiplication**.

Quotient is the result of **division**.

Quotient is probably the scariest word of the bunch. Luckily for you, it also is the one that comes up least often.

As always when translating, start by assigning variables. Let the number be n. "The product of a number and 5" means the result of multiplying n and 5. You can show multiplication with a multiplication sign, parentheses, or simply by writing the number in front of the variable:

$$n \times 5 \qquad (n)(5) \qquad 5n$$

All three of these expressions represent the product of a number and 5. The last is simplest, though, so use that.

The next part of the statement is "is equal to." This is easy; it's just an equal sign (=).

"The sum of that number and 12" means the result of adding n and 12. This translates to: $n + 12$.

Putting it all together, you get $5n = n + 12$.

WISE POINTS

★ Order doesn't matter with addition and multiplication: $6 + 2 = 2 + 6$, and $6 \times 2 = 2 \times 6$.

★ Order is VERY important with subtraction and division: $6 - 2 \neq 2 - 6$, and $6 \div 2 \neq 2 \div 6$.

Solve: Now it's time to solve. You have one equation with one variable. The basic rule of thumb is that you need as many different equations with the variables as you have variables you want to solve for. Since you've got one variable and one equation, you're good to go.

To solve for an equation, you isolate the variable on one side. That means you get it all by itself on one side of the equation. And how do you do that? Well, you "undo" everything that has been done to it. To undo an operation, use the opposite arithmetic operation. As long as you do the same thing to both sides of the equation, you keep the equality.

Isolation may be good for solving equations, but I am not a fan of it. My parents grounded me for a week once—no phone and no computer. Talk about isolation!

In this equation you have variables on both sides: $5n$ on the left, and n is added to 12 on the right. The opposite of addition is subtraction, so subtract n from both sides:

$$5n = n + 12$$
$$5n - n = n + 12 - n$$
$$4n = 12$$

Now you have all the n terms on one side. Since $4n$ means that n has been multiplied by 4, you must undo this to be left with plain old n. The opposite of multiplication is division, so divide both sides by 4:

$$4n = 12$$
$$\frac{4n}{4} = \frac{12}{4}$$
$$n = 3$$

WISE POINTS

* Subtraction undoes addition and addition undoes subtraction.
* Division undoes multiplication and multiplication undoes division.

Check Work: To check your work, try your number in the original word problem. If it works, you know that you're right. So, the original problem says that the product of a number and 5 is equal to the sum of that number and 12. The product of 3 and 5 is (3)(5) = 15, and the sum of 3 and 12 is 3 + 12 = 15. Since 15 = 15, you know that 3 is your number.

WISE NOTE

Some word problems about numbers will just ask you to set up an equation that could be used to solve for the unknown. Some will ask you to actually find the unknown. If you're given a multiple-choice question that asks you to actually find the unknown, there's another way to do it: You can work backward from the answer choices to see which is correct. You do this the same way that you check an answer: plug the number into the original word problem to see if it works.

Get Wise!

Think you got it? Then try the following word problems about numbers.

1. Two more than three times a number is 20. What is the number?

    (A) 4   **(D)** $12\frac{1}{2}$

    (B) 6   **(E)** 18

    (C) $7\frac{1}{3}$

2. Sixteen decreased by a number is equal to the product of that number and 3. What is the number?

    (A) 4   **(D)** 12

    (B) 8   **(E)** 13

    (C) 11

3. One-third of a number is equal to 24 less than that number. What is the number?

    (A) 8   **(D)** 24

    (B) 12   **(E)** 36

    (C) 18

4. Two more than a number divided by 2 is 2 less than the number. What is the number?

    (A) 0   **(D)** 6

    (B) 2   **(E)** 8

    (C) 4

5. Four more than a number is 3 less than twice that number. What is the number?

(A) 0.5   (D) 14

(B) 3.5   (E) 18

(C) 7

How Wise?

Check your answers on page 47.

ALL IN A ROW

Some number problems will talk about consecutive integers. "Consecutive" basically means in a row, without any gaps. It's important to know how to represent consecutive numbers algebraically so you can set up these equations.

I wouldn't want this getting out or anything, but I was the Ping-Pong champ at summer camp for three consecutive years.

Example: The sum of three consecutive integers is equal to the greatest of these integers plus 15. What is the second of these integers?

Solution: Translate the statement to a math equation, and then solve the equation. If you have time, it's a good idea to check your work.

Translate: Use one variable to represent the consecutive integers. If you set up an equation with three different variables, you won't be able to solve for their values. Remember, you need as many different equations including your variables as there are variables in order to get a solution.

So how do you represent these three numbers with one variable? Let the first consecutive integer be x. The second integer follows right after the first, so it must be one greater. Therefore, it is $x + 1$. The third integer is one greater than the second, so it is $x + 1 + 1 = x + 2$.

That means that "the sum of three consecutive integers" translates to $x + x + 1 + x + 2$.

"Is equal to" translates to = . "The greatest of these integers plus 15" translates to $x + 2 + 15$. Putting it all together, you get $x + x + 1 + x + 2 = x + 2 + 15$.

Solve: Now that you have your equation, you can solve for the value of x. The first step is to simplify each side of the equation by combining like terms.

$$x + x + 1 + x + 2 = x + 2 + 15$$

$$3x + 3 = x + 17$$

Now do the same thing to both sides until you isolate the variable.

$$3x + 3 = x + 17$$
$$3x + 3 - x = x + 17 - x$$
$$2x + 3 = 17$$
$$2x + 3 - 3 = 17 - 3$$
$$2x = 14$$
$$\frac{2x}{2} = \frac{14}{2}$$
$$x = 7$$

Be careful! You've found the value of the variable, but you're not done yet. You set up the equation so that x was the smallest of the three integers but the question asks for the second. That means that the second integer is

$$x + 1 = 7 + 1 = 8.$$

Check Work: Since the first integer is 7 and the second is 8, the third must be 9. Check to see whether these values work in the original word problem: "the sum of three consecutive integers is equal to the greatest of these integers plus 15." The sum of these three integers is $7 + 8 + 9 = 24$. The greatest integer plus 15 is $9 + 15 = 24$. Since $24 = 24$, you know that your solution is correct.

WISE NOTE

Remember, solving for the variable doesn't always give you the answer the question asks for. You might need to use the value of the variable to find another value. Always reread the question to make sure that you're answering the question that's asked.

Example: The sum of seven consecutive odd integers is 119. What is the greatest of these integers?

Solution: You could tackle this problem head-on by setting up an equation and solving for the value of a variable. We'll show you how to set up the problem this way, but then we'll give you a quicker, easier way to solve.

Cool! A shortcut. That's just what I'm looking for in this book.

Standard Setup: To write an equation, you'd let the first integer be x as before. But now you're dealing with consecutive odd integers, not plain old consecutive integers. So how do you express them? Since regular consecutive integers are 1 apart, you add increments of 1 to the variable to express them: x, $x + 1$, $x + 2$, $x + 3$, and so on. How far apart are consecutive odd integers? Well, 1 and 3 are 2 apart, as are 3 and 5, and 5 and 7. Since consecutive odd integers are 2 apart, you add increments of 2: x, $x + 2$, $x + 4$, $x + 6$, and so on. You need to write an expression for each of the seven integers and then sum them to 119. Then you could solve this equation for x, which you'd then plug into the expression for the greatest integer to find its value. This is all pretty straightforward but also rather time-consuming. And nothing personal, but the more work you do, the more room you'll have to make a careless mistake.

Shortcut: Want to learn the quicker way? There's a special fact about consecutive integers. Their average is equal to the middle term in the group. You know that there are seven integers and that their sum is 119. Since average is

$\dfrac{\text{Sum of terms}}{\text{Number of terms}}$, the average of these integers is $\dfrac{119}{7} = 17$. There are seven integers in this group so the middle one is the fourth. If the fourth is 17, then the fifth is 19, the sixth is 21, and the seventh and greatest is 23. Ta da—you're done!

WISE POINT

You can also find the average of a set of consecutive integers by averaging the first and last terms.

Check Work: As always, check your work if you have time. If the four greatest numbers are 17, 19, 21, and 23, the first three must be 11, 13, and 15. These add up to 11 + 13 + 15 + 17 + 19 + 21 + 23 = 119, so you've got it.

WISE NOTE

To figure out how to express consecutive integers in terms of one variable, figure out how far apart the integers are. Even integers are 2 apart, so you express them by adding increments of 2 to the variable that stands for the first integer (as you do with odd integers). Consecutive multiples of 5 are 5 apart, so you express those by adding increments of 5 to the original variable, and so on.

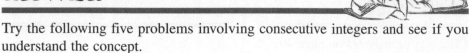

Get Wise!

Try the following five problems involving consecutive integers and see if you understand the concept.

1. The sum of three consecutive even integers is 8 more than the middle integer. What is the value of the greatest integer?

   (A) 2   (D) 6

   (B) 3   (E) 12

   (C) 4

2. The sum of four consecutive multiples of 5 is equal to the product of the first integer and 7. What is the first integer?

   (A) 5   (D) 20

   (B) 10   (E) 25

   (C) 15

3. The average of the eleven consecutive odd integers is 17. What is the first of these integers?

   (A) 1   (D) 7

   (B) 3   (E) 9

   (C) 5

4. What is the average of the first twenty-five positive integers?

   (A) 5   (D) 24

   (B) 12.5   (E) 325

   (C) 13

5. The product of two consecutive positive integers is equal to six times the first integer. What is the greater of these two integers?

   (A) 3   (D) 12

   (B) 5   (E) 30

   (C) 6

How Wise?

Check your answers on page 49.

EXTRA! EXTRA!

There are some other words and tips you need to know to help you decode equations.

Decoding? I feel like one of the Spy Kids.

Example: Six times the reciprocal of a number is equal to that number decreased by 1. What is the number?

Solution: Translate the statement to a math equation, and then solve the equation. If you have time, it's a good idea to check your work.

Translate: The word you need to know here is "reciprocal." A math text-book defines it as the number that, when multiplied by another number, has a product of 1. We'll define it as a number with its numerator and denominator flipped. Any integer can be thought of as a fraction with a denominator of 1. So 4 could be written as $\frac{4}{1}$, which makes it clear that flipping its numerator and denominator gives you $\frac{1}{4}$. So, the reciprocal of 4 is $\frac{1}{4}$. Likewise, the reciprocal of $\frac{1}{4}$ is 4. (Check it out: $\frac{4}{1} \times \frac{1}{4} = 1$. This is pretty simple, don't you think?)

Okay, back to our example. Let the number be n. That makes its reciprocal $\frac{1}{n}$. "Six times the reciprocal of a number" translates to $6 \times \frac{1}{n} = \frac{6}{n}$. "Is equal" is clearly an equal sign, = . "That number decreased by 1" translates to $n - 1$. Putting it all together gives you the equation $\frac{6}{n} = n - 1$.

Solve: After translating, you'll probably be horrified to see that you have a fraction on one side of the equation. Don't worry. There are two ways to get rid of it. One is to cross-multiply the equation; the other is to multiply everything in the equation by the denominator of the fraction. Either way you end up with the same equation.

$$\frac{6}{n} = n - 1$$
$$\frac{6}{n} = \frac{n-1}{1}$$
$$6(1) = n(n-1)$$
$$6 = n^2 - n$$

$$\frac{6}{n} = n - 1$$
$$n\left(\frac{6}{n}\right) = n(n-1)$$
$$6 = n^2 - n$$

Now that you've gotten rid of the fraction, you probably won't be happy to realize that you've got a quadratic equation on your hands. Don't worry—they're not nearly as bad as you fear. And just so you don't worry too much, we'll let you know right now that this equation can be factored. That means you won't need to mess with the quadratic formula.

It's gone from bad to worse! The thought of quadratic equations gives me hives! Maybe the quadratic formula is some kind of antidote for that.

The first step when working with a quadratic equation is to put it in standard form: $ax^2 + bx + c = 0$.

This looks kind of scary and technical, but all you have to do is follow these steps:

Put the squared variable first.

Put the variable raised to the first power next.

Follow with the constant (the term without a variable).

Set all of it equal to 0.

You now have terms on both sides of the equation. This means that the first step is to get all of them on one side. You can do this by subtracting 6 from both sides.

$$6 = n^2 - n$$
$$6 - 6 = n^2 - n - 6$$
$$0 = n^2 - n - 6$$
$$n^2 - n - 6 = 0$$

Since the quantities on both sides of the equal sign are equal, you can flip them so that you have standard form. Now you're ready to solve. To factor, you use FOIL in reverse. FOIL is the order in which you multiply binomial terms to come up with a quadratic product.

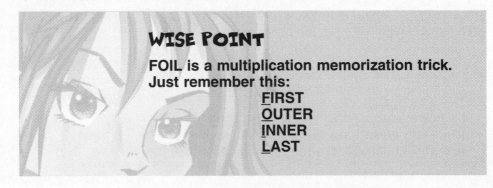

WISE POINT

FOIL is a multiplication memorization trick. Just remember this:

FIRST
OUTER
INNER
LAST

Start by setting up two parentheses. Since you want to end up with an n^2, the first term in each should be n.

$$(n+\quad)(n+\quad) = 0$$

You know that the last two numbers will multiply together to give you –6. Since a positive times a negative is negative, one of the numbers will be positive and one negative. The numbers could be 2 and 3 or 6 and 1. The key here is to see which will give you a middle term of $-x$. Using –3 and 2 works:

$$(n-3)(n+2) = 0$$

Set each factor equal to 0 and you get $n = 3$ or $n = -2$.

Check Work: You have two answers for the number so check both to see if they work in the original question. "Six times the reciprocal of a number is equal to that number decreased by 1."

$$6 \times \frac{1}{3} = 3 - 1 \qquad\qquad 6 \times \left(-\frac{1}{2}\right) = -2 - 1$$
$$2 = 2 \qquad\qquad\qquad\qquad -3 = -3$$

Both work, so the number could be either 3 or –2. But careful! A question like this could specify that they are looking only for the positive number, or only for the negative one. You don't want to go through all this work and then get it wrong in the end by not reading closely enough.

Move over, Juliet—I'd be a major tragic figure if I did all that work and then blew it at the last minute.

Get Wise!

Now it's your turn. Try the following extra number problems.

1. The reciprocal of a number is equal to $\frac{1}{9}$ of that number. If the number is positive, what is its value?

(A) $\frac{1}{3}$

(B) 3

(C) 6

(D) 9

(E) 81

2. Eight times the reciprocal of a number minus 2 is equal to 2. What is the number?

(A) 0

(B) 2

(C) 4

(D) 6

(E) 8

3. A number squared plus the number is equal to five times the number. If the number is positive, what is its value?

(A) 4

(B) 5

(C) 6

(D) 9

(E) 20

I must be losing it from all these number problems. This one sounds just as complicated as the start of the Gettysburg Address.

4. Nine less than a number squared is equal to 16. If this number is positive, what is its value?

(A) 3   (D) 6

(B) 4   (E) 7

(C) 5

5. Six times the reciprocal of a number is equal to 24. What is the number?

(A) 12   (D) $\dfrac{1}{4}$

(B) 6   (E) $\dfrac{1}{6}$

(C) 4

How Wise?

Check your answers on page 51.

A Word to the Wise

* Know the words that stand for the results of the four basic arithmetic operations.

* Solve an equation by doing the same thing to both sides until the variable is isolated.

* Addition and subtraction undo each other; multiplication and division undo each other.

* To find the reciprocal of a number, flip its numerator and denominator.

* To get rid of a fraction in an equation, multiply both sides of the equation by the denominator of the fraction.

* To solve a quadratic equation, put it in standard form and factor it.

* To represent consecutive integers with one variable, figure out how far apart they are and add to the original variable in increments of that amount.

* The average of consecutive integers is equal to the middle integer.

* Always read a word problem carefully to make sure that you're answering the question that's asked.

ANSWERS TO CHAPTER 2: PRACTICE EXERCISES

Word Problems About Numbers (Page 34)

1. **The correct answer is (B).** Let the number be n. "Three times a number" translates to $3n$, so 2 more than this is $3n + 2$. "Is 20" translates to $= 20$. Putting it together gives you $3n + 2 = 20$. Isolate n to solve for its value:

$$3n + 2 = 20$$
$$3n = 18$$
$$n = 6$$

2. **The correct answer is (A).** Let the number be n. "Sixteen decreased by a number" translates to $16 - n$. "Is equal to the product of that number and 3" translates to $= 3n$. Putting it all together gives you $16 - n = 3n$. Isolate n to solve for its value:

$$16 - n = 3n$$
$$16 = 4n$$
$$4 = n$$

3. **The correct answer is (E).** Let the number be n. "One-third of a number" translates to $\frac{1}{3} \times n$, which simplifies to $\frac{n}{3}$. "Is equal to 24 less than that number" $= n - 24$. Putting it all together gives you $\frac{n}{3} = n - 24$. Isolate n to solve for its value:

$$\frac{n}{3} = n - 24$$
$$n = 3(n - 24)$$
$$n = 3n - 72$$
$$-2n = -72$$
$$n = 36$$

4. **The correct answer is (E).** Let the number be n. "A number divided by 2" translates to $\dfrac{n}{2}$, so 2 more than this is $\dfrac{n}{2}+2$. "Is 2 less than the number" translates to $= n - 2$. Putting it all together gives you $\dfrac{n}{2}+2 = n-2$. Isolate n to solve for its value:

$$\frac{n}{2}+2 = n-2$$
$$n+4 = 2(n-2)$$
$$n+4 = 2n-4$$
$$n+8 = 2n$$
$$8 = n$$

5. **The correct answer is (C).** Let the number be n. "Four more than a number" translates to $n + 4$. "Twice that number" translates to $2n$, so 3 less than that is $2n - 3$. Putting it all together gives you $n + 4 = 2n - 3$. Isolate n to solve for its value:

$$n+4 = 2n-3$$
$$n+7 = 2n$$
$$7 = n$$

Consecutive Integers (Page 39)

1. **The correct answer is (D).** Let the first even integer be n. Since even integers are 2 apart, represent the second and third integers by adding increments of 2 to n: $n + 2$ and $n + 4$. "The sum of three consecutive even integers" translates to $n + n + 2 + n + 4$. "Is 8 more than the middle integer" translates to $= 8 + n + 2$. Putting it all together gives you $n + n + 2 + n + 4 = 8 + n + 2$. Isolate n to solve for its value:

$$n + n + 2 + n + 4 = 8 + n + 2$$
$$3n + 6 = n + 10$$
$$3n = n + 4$$
$$2n = 4$$
$$n = 2$$

This is the smallest integer. The greatest is $n + 4 = 2 + 4 = 6$.

2. **The correct answer is (B).** Let the first multiple of 5 be n. Since multiples of 5 are 5 apart, represent the second, third, and fourth multiples by adding increments of 5 to n: $n + 5$, $n + 10$, and $n + 15$. "The sum of four consecutive multiples of 5" translates to $n + n + 5 + n + 10 + n + 15$. "Is equal to the product of the first integer and 7" translates to $= 7 \times (n)$, or $7n$. Putting it all together gives you $n + n + 5 + n + 10 + n + 15 = 7n$. Isolate n to solve for its value:

$$n + n + 5 + n + 10 + n + 15 = 7n$$
$$4n + 30 = 7n$$
$$30 = 3n$$
$$10 = n$$

This is the first consecutive multiple of 5, which is what the question asked for.

3. **The correct answer is (D).** You could represent the consecutive integers in terms of one variable, set up an equation, and solve for its value. But there's a quicker way. The average of consecutive integers is equal to the middle term. You know that the average is 17, so this is the middle term. Since there are eleven integers, this is the sixth integer. So the fifth is 15, the fourth is 13, the third is 11, the second is 9, and the first is 7.

4. **The correct answer is (C).** You could represent the consecutive integers in terms of one variable, set up an equation, and solve for its value. But there's a quicker way. The average of consecutive integers is equal to the middle term. You're asked for the average of the first twenty-five positive integers, or 1 through 25. This is a total of twenty-five terms, so the middle one is 13. That means that the average of the first twenty-five positive integers is 13.

5. **The correct answer is (C).** Let the first integer be n. Since consecutive integers are 1 apart, represent the second integer by 1 added to n: $n + 1$. "The product of two consecutive positive integers" translates to $n(n + 1)$. "Is equal to six times the first integer" translates to $= 6n$. Putting it all together gives you $n(n + 1) = 6n$. Isolate n to solve for its value:

$$n(n+1) = 6n$$
$$n^2 + n = 6n$$
$$n^2 - 5n = 0$$
$$n(n-5) = 0$$

$$n = 0 \qquad\qquad n - 5 = 0$$
$$n = 5$$

The question says that the integers are both positive. Since 0 is neither positive nor negative, it cannot be part of the correct answer. That means the first integer must be 5 and the second is 6.

Extra Number Problems (Page 44)

1. **The correct answer is (B).** Let the number be n. "The reciprocal of a number" translates to $\dfrac{1}{n}$. "Is equal to $\dfrac{1}{9}$ of that number" translates to $\dfrac{1}{9} \times n$ or $\dfrac{n}{9}$. Putting it all together gives you $\dfrac{1}{n} = \dfrac{n}{9}$. Isolate n to solve:

$$\frac{1}{n} = \frac{n}{9}$$
$$n^2 = 9$$
$$\sqrt{n^2} = \sqrt{9}$$

$\sqrt{9}$ is either 3 or –3, so n must be either 3 or –3. The question says the number is positive, so the correct answer is $n = 3$.

2. **The correct answer is (B).** Let the number be n. "The reciprocal of a number" translates to $\dfrac{1}{n}$, so eight times this reciprocal minus 2 is $8 \times \dfrac{1}{n} - 2$ or $\dfrac{8}{n} - 2$. "Is equal to 2" translates to = 2. Putting it all together gives you $\dfrac{8}{n} - 2 = 2$. Isolate n to solve:

$$\frac{8}{n} - 2 = 2$$
$$\frac{8}{n} = 4$$
$$4n = 8$$
$$n = 2$$

3. **The correct answer is (A).** Let the number be n. "A number squared" translates to n^2, so "a number squared plus the number" gives you $n^2 + n$. "Is equal to five times the number" translates to $= 5n$. Putting it all together gives you $n^2 + n = 5n$. Isolate n to solve:

$$n^2 + n = 5n$$
$$n^2 - 4n = 0$$
$$n(n-4) = 0$$

$$n = 0 \qquad\qquad n - 4 = 0$$
$$n = 4$$

The question says the number is positive, so the answer is 4.

4. **The correct answer is (C).** Let the number be n. "A number squared" translates to n^2, so 9 less than that is $n^2 - 9$. "Is equal to 16" translates to $= 16$. Putting it all together gives you $n^2 - 9 = 16$. Isolate n to solve:

$$n^2 - 9 = 16$$
$$n^2 = 25$$

Since $n^2 = 25$, n must be either 5 or –5. The question says the number is positive, so the answer is 5.

5. **The correct answer is (D).** Let the number be n. "The reciprocal of a number" translates to $\dfrac{1}{n}$, so six times this is $6 \times \dfrac{1}{n}$ or $\dfrac{6}{n}$. "Is equal to 24" translates to $= 24$. Putting it all together gives you $\dfrac{6}{n} = 24$.

Isolate n to solve:

$$\frac{6}{n} = 24$$
$$24n = 6$$
$$n = \frac{6}{24}$$
$$n = \frac{1}{4}$$

A Fraction of a Number

UNDERSTANDING WORD PROBLEMS WITH FRACTIONS

Fractions and word problems together—my favorite! Does it get any better than this? (You know I'm being sarcastic, right?)

While fractions may seem like an extra wrinkle in a problem, you use them all the time, probably without even thinking. When you say that your team won two-thirds of the games it played this year, or figure out how much flour you need if you decide to cut a cookie recipe in half, you've entered the wonderful world of word problems with fractions.

The World of Word Problems with Fractions—that's a great new attraction at EPCOT, isn't it?

Some problems will describe a situation with actual numbers and ask you to come up with the fraction it represents. That's it. All you really have to remember is that the number at the top of the fraction (the numerator) is the part that's being talked about, and the number at the bottom of the fraction (the denominator) is the whole.

WISE POINTS

* The top of a fraction is called the numerator.
* The bottom of a fraction is called the denominator.

Example: A band is coming to town for a sold-out concert. Twelve members of its local fan club were able to get tickets, but the remaining 28 members were not. What fractional part of the fan club is lucky enough to be going to the big show?

Solution: Put the part over the whole. In this question, the part is the number of people who actually got tickets. That's 12. The whole is the total number of people in the fan club. This is the number of people who got tickets plus the number of people who did not. That's 12 + 28 = 40. So the fractional part of the fan club going to the concert is $\frac{12}{40}$, or $\frac{3}{10}$.

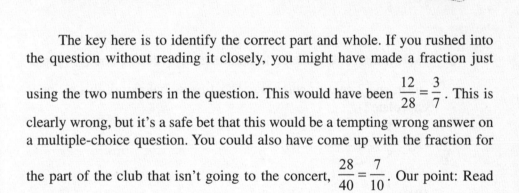

Only $\frac{3}{10}$ are going to the concert? Some fan club!

The key here is to identify the correct part and whole. If you rushed into the question without reading it closely, you might have made a fraction just using the two numbers in the question. This would have been $\frac{12}{28} = \frac{3}{7}$. This is clearly wrong, but it's a safe bet that this would be a tempting wrong answer on a multiple-choice question. You could also have come up with the fraction for the part of the club that isn't going to the concert, $\frac{28}{40} = \frac{7}{10}$. Our point: Read carefully.

Get Wise!

Think you got it? Then try the following word problems with fractions.

1. There are 25 dogs and 15 cats at the local animal shelter. What part of the animals at the shelter are cats?

(A) $\dfrac{1}{5}$ (D) $\dfrac{5}{8}$

(B) $\dfrac{3}{8}$ (E) $\dfrac{5}{2}$

(C) $\dfrac{3}{5}$

2. A brownie recipe calls for 1 cup of butter, $\dfrac{1}{2}$ cup of flour, $\dfrac{1}{2}$ cup of sugar, and 2 cups of chocolate. What fractional part of this recipe is chocolate?

(A) $\dfrac{1}{8}$ (D) $\dfrac{1}{2}$

(B) $\dfrac{1}{4}$ (E) 2

(C) $\dfrac{1}{3}$

3. Sharon has a total of 15 cousins. If 6 of them are boys, what fractional part of her cousins are girls?

(A) $\dfrac{1}{5}$ (D) $\dfrac{5}{3}$

(B) $\dfrac{2}{5}$ (E) $\dfrac{5}{2}$

(C) $\dfrac{3}{5}$

4. There are 12 blue cars, 9 red cars, and 15 gray cars in a certain parking lot. What fractional part of the cars in this lot is NOT blue?

(A) $\dfrac{1}{4}$ (D) $\dfrac{2}{3}$

(B) $\dfrac{1}{3}$ (E) $\dfrac{5}{6}$

(C) $\dfrac{1}{2}$

5. Jake got $150 for his birthday. He spends $25 on a video game and $35 on some CDs. What fractional part of his money does he have left after this?

(A) $\dfrac{1}{6}$ (D) $\dfrac{3}{5}$

(B) $\dfrac{7}{30}$ (E) $\dfrac{2}{3}$

(C) $\dfrac{2}{5}$

How Wise?

Check your answers on page 71.

A FRACTION OF A FRACTION

Some questions ask about a fraction of a fraction. Before that makes you too dizzy, we'll tell you what it means. Say your team won $\frac{2}{3}$ of the soccer games it played this year. Of those wins, $\frac{1}{4}$ were at home. So how many soccer games did you win at home this year? Since "of" translates to multiplication, you multiply the two fractions together. This means that your team won $\frac{2}{3} \times \frac{1}{4} = \frac{2}{12} = \frac{1}{6}$ of the games it played at home.

That doesn't sound like much of a home-field advantage to me.

Example: On a certain sitcom, $\frac{1}{3}$ of the cast members are blonds. If $\frac{3}{4}$ of the blonds are women, what fractional part of the cast members on the sitcom are blond men?

Solution: The question is talking about a fraction of a fraction, so you know that you'll have to multiply. But don't just rush in and multiply $\frac{1}{3}$ by $\frac{3}{4}$. Read the question carefully to see what it's asking for. It wants to know what fractional part of the cast are blond men. This means that you need to multiply the fraction of the cast that is blond by the fraction of blonds that are men.

The fraction $\frac{1}{3}$ represents the fraction of blonds. However, the fraction of $\frac{3}{4}$ represents the fraction of blond women. If you multiplied these two together you'd come up with the fractional part of the cast that are blond women. You need to know what fraction of the blond cast members are men. Since $\frac{3}{4}$ of them are women, the remaining $1 - \frac{3}{4} = \frac{1}{4}$ must be men.

So the fraction of the cast members that are blond men is $\frac{1}{3} \times \frac{1}{4} = \frac{1}{12}$.

In some of these problems, a fraction is implied instead of directly stated. You're given information about the numbers involved, but you need to put the fraction together yourself.

Oh, I know what "implied" means. That's when you want to tell someone something but you can't quite come out and say it. My friend Marisa wanted to buy this awful shirt until I implied that it would look better on her mother.

Example: There are twice as many girls as there are boys in a school's chess club. If $\frac{3}{8}$ of the boys are seniors, they make up what fractional part of the club?

Solution: You want the fractional part of the chess club that consists of senior boys, so you need to multiply the fraction of boys in the club by the fraction of boys that are seniors. You know that $\frac{3}{8}$ of the boys are seniors, but you aren't told what fraction of the club are boys.

Use the information you are given to write the implied fraction. You're told that there are twice as many girls in the club as there are boys. That means that for every 2 girls there is 1 boy, making the fraction of boys in the club $\frac{1}{2+1} = \frac{1}{3}$. Be careful here; an easy mistake to make is to incorrectly translate this fraction as $\frac{1}{2}$.

So, $\frac{1}{3} \times \frac{3}{8} = \frac{3}{24}$, or $\frac{1}{8}$, of the club are senior boys.

Get Wise!

Now's your chance to test yourself by trying the following problems involving fractions of fractions.

1. Drew saves $\frac{1}{4}$ of his paycheck each week. He spends $\frac{1}{2}$ of the remainder on going out with friends. What fractional part of his weekly paycheck does Drew spend on going out with friends?

(A) $\frac{1}{8}$

(D) $\frac{3}{8}$

(B) $\frac{1}{6}$

(E) $\frac{3}{4}$

(C) $\frac{1}{4}$

I wish more of my friends spent that fraction of their money hanging out with me. My allowance never lasts the week.

2. One-sixth of the crayons in a box are pastels. If $\frac{1}{4}$ of the pastel crayons are broken, what fractional part of the box of crayons are broken pastels?

(A) $\frac{1}{24}$

(D) $\frac{5}{12}$

(B) $\frac{1}{18}$

(E) $\frac{5}{8}$

(C) $\frac{5}{24}$

3. There are three times as many chocolate cupcakes as vanilla ones. If $\frac{1}{4}$ of the chocolate cupcakes have sprinkles, what fractional part of the cupcakes are chocolate with sprinkles?

(A) $\frac{1}{16}$ (D) $\frac{3}{16}$

(B) $\frac{1}{12}$ (E) $\frac{3}{4}$

(C) $\frac{1}{6}$

4. A test is made up of multiple-choice and open-ended questions, and $\frac{1}{5}$ of the questions are open-ended. If Inez answers $\frac{7}{8}$ of the multiple-choice questions correctly, what fractional part of the test does this represent?

(A) $\frac{1}{40}$ (D) $\frac{8}{13}$

(B) $\frac{1}{10}$ (E) $\frac{7}{10}$

(C) $\frac{7}{40}$

5. A store sells twice as many DVDs as VHS tapes. If $\frac{2}{5}$ of the VHS tapes they sell are action films, what fractional part of the store's video sales are VHS action films?

(A) $\frac{2}{15}$ (D) $\frac{3}{10}$

(B) $\frac{1}{5}$ (E) $\frac{2}{5}$

(C) $\frac{4}{15}$

How Wise?

Check your answers on page 72.

USING FRACTIONS TO FIND THE WHOLE

Suppose that your friend is throwing a party. She tells you that $\frac{2}{3}$ of the people she invited are coming, which is 20 kids. You're wondering how many people her parents let her invite in the first place. How do you find out without asking her?

When I throw a party, there are no fractions about it. All of the people who are lucky enough to be invited come.

You'll need to do a little translating here. You want to know the original number of people invited, so let that be x (or n, or whatever variable you want). You know that $\frac{2}{3}$ of that number are coming; since "of" means multiply, you translate this as $\frac{2}{3}x$. This is 20 kids, so you translate as = 20. Put it all together and you can solve for x:

$$\frac{2}{3}x = 20$$
$$2x = 60$$
$$x = 30$$

Example: Jimmy's mom works at the department store and gets a discount off purchases she makes there. Her discount means that she pays $\frac{4}{5}$ of the ticket price for an item. If she bought a sweater for $40, what was its ticket price? (For your convenience, there is no sales tax involved.)

Solution: Set up an equation. You want to find the ticket price of the sweater, so let that be x. Jimmy's mom pays $\frac{4}{5}$ of the ticket price, which translates to $\frac{4}{5}x$. The price she paid is $40, which translates to = $40. Put this all together into an equation and solve for x:

$$\frac{4}{5}x = \$40$$
$$4x = 5(\$40)$$
$$4x = \$200$$
$$x = \$50$$

I wish my mom could get me a discount on clothes and stuff. She's a dentist, so all I ever get is flavored floss and extra toothbrushes.

WISE NOTE

You know something about the size of the right answer before you even set up your equation. Since Jimmy's mom pays just a fraction of the ticket price, you know that the ticket price must be more than what she paid. If this were a multiple-choice question, you could have eliminated any choices that were less than $40.

Get Wise!

On to the next batch of questions. This time, you'll have to work with the sum, using fractions to find the whole.

1. Fourteen students show up for soccer practice on Thursday. If this

   represents $\frac{7}{8}$ of the soccer team, what is the total number of students

   on the team?

   (A) 12   (D) 20

   (B) 15   (E) 28

   (C) 16

2. Mr. Big puts $9,000 in charges on his credit card. If these charges are

   equal to $\frac{3}{4}$ of his credit limit, what is his credit limit?

   (A) $6,750   (D) $27,000

   (B) $12,000   (E) $36,000

   (C) $18,000

Mr. Big sure is going to have a big credit card bill!

3. This month, Brenna has used 260 minutes on her cell phone. If she

   has $\frac{12}{25}$ of the minutes on her monthly plan limit left, what is the

   total number of minutes in her monthly plan limit?

   (A) 272   (D) 540

   (B) 500   (E) 600

   (C) 520

I went way over my minute limit once, and, boy, that was expensive! My parents flipped—I didn't get my

allowance for the next $\frac{1}{4}$ of the year!

4. Greg has an MP3 player. He only has room for 600 more minutes of music left on it. If he has filled it to $\frac{4}{5}$ of its capacity, what is the total number of minutes of music that it can hold?

(A) 480   (D) 2,400

(B) 1,200   (E) 3,000

(C) 1,800

5. Henry received four times as many votes as his only opponent in the school election. If his opponent received 40 votes, what is the total number of votes that were cast?

(A) 50   (D) 160

(B) 80   (E) 200

(C) 120

How Wise?

Check your answers on page 73.

A Word to the Wise

★ Remember, a fraction represents some part over a whole: $\dfrac{\text{part}}{\text{whole}}$

★ To find a fraction of a fraction, multiply them together.

★ If a question includes an implied fraction, use the information from the question to put it together.

★ To find a whole from a fraction, translate into an equation.

ANSWERS TO CHAPTER 3: PRACTICE EXERCISES

Understanding Word Problems with Fractions (Page 58)

1. **The correct answer is (B).** You want to know what fractional part of the animals are cats, so the number of cats is the part: 15. There are 15 cats and 25 dogs, so their sum is the whole: $15 + 25 = 40$. So cats are $\dfrac{15}{40} = \dfrac{3}{8}$ of the animals at the shelter.

2. **The correct answer is (D).** You want to know what fractional part of the recipe is chocolate, so the amount of chocolate is the part: 2 cups. In the recipe there are 1 cup of butter, $\dfrac{1}{2}$ cup flour, $\dfrac{1}{2}$ cup butter, and 2 cups of chocolate, so their sum is the whole: $1 + \dfrac{1}{2} + \dfrac{1}{2} + 2 = 4$ cups. So, chocolate is $\dfrac{2}{4} = \dfrac{1}{2}$ of the recipe.

3. **The correct answer is (C).** You want to know what fractional part of Sharon's cousins are girls, so the number of girls is the part. You aren't given this number but you can find it. Since Sharon has a total of 15 cousins and 6 of them are boys, the remaining $15 - 6 = 9$ are girls. Sharon's total number of cousins is the whole: 15. So $\dfrac{9}{15} = \dfrac{3}{5}$ of Sharon's cousins are girls.

4. **The correct answer is (D).** You want to know what fractional part of the cars is NOT blue. This means that the part is equal to the sum of the red and gray cars: $9 + 15 = 24$. There are 12 blue cars, 9 red cars, and 15 gray cars, so their sum is the whole: $12 + 9 + 15 = 36$. So $\dfrac{24}{36} = \dfrac{2}{3}$ of the cars in the lot are NOT blue.

5. **The correct answer is (D).** You want to know what fractional part of his birthday money Jake has left, so this amount is the part. You're not given the amount he has left but you can find it. Since Jake received a total of $150 and spent $25 + $35, $150 – ($25 + $35) = $150 – $60 = $90 is left. The total amount Jake received is the whole: $150. So Jake has $\dfrac{\$90}{\$150} = \dfrac{3}{5}$ of his birthday money left.

Fractions of a Fraction (Page 63)

1. **The correct answer is (D).** Drew saves one-fourth of his paycheck each week, so he spends $1 - \dfrac{1}{4} = \dfrac{3}{4}$ of it. Drew spends one-half of this remainder on going out with friends. So he spends $\dfrac{3}{4} \times \dfrac{1}{2} = \dfrac{3}{8}$ of his paycheck going out with friends.

2. **The correct answer is (A).** Since $\dfrac{1}{6}$ of the crayons are pastels, and $\dfrac{1}{4}$ of the pastels are broken, $\dfrac{1}{6} \times \dfrac{1}{4} = \dfrac{1}{24}$ of the crayons in the box are broken pastels.

3. **The correct answer is (D).** There are three times as many chocolate cupcakes as vanilla ones, so there are 3 chocolate cupcakes for every 1 vanilla. This means that $\dfrac{3}{3+1} = \dfrac{3}{4}$ of the cupcakes are chocolate. Since $\dfrac{1}{4}$ of the chocolate cupcakes have sprinkles, $\dfrac{3}{4} \times \dfrac{1}{4} = \dfrac{3}{16}$ of the cupcakes are chocolate with sprinkles.

4. **The correct answer is (E).** Since $\frac{1}{5}$ of the questions on the test are

open-ended, the remaining $1 - \frac{1}{5} = \frac{4}{5}$ of them must be multiple choice.

Inez answers $\frac{7}{8}$ of the multiple-choice questions correctly. So this

represents $\frac{4}{5} \times \frac{7}{8} = \frac{28}{40}$, or $\frac{7}{10}$, of the test.

5. **The correct answer is (A).** The store sells twice as many DVDs as VHS tapes, so it sells 2 DVDs for every 1 VHS tape. This means that

$\frac{1}{2+1} = \frac{1}{3}$ of the videos sold are VHS tapes. Since $\frac{2}{5}$ of the VHS

tapes sold are action films, $\frac{1}{3} \times \frac{2}{5} = \frac{2}{15}$ of the videos that the store

sells are VHS tapes of action films.

Using Fractions to Find the Whole (Page 68)

1. **The correct answer is (C).** Let the total number of students on the

soccer team be x. Since $\frac{7}{8}$ of the team shows up for practice, this

translates to $\frac{7}{8}x$. This is equal to 14 students, so your equation is

$\frac{7}{8}x = 14$. Solve for x:

$$\frac{7}{8}x = 14$$
$$7x = 8(14)$$
$$7x = 112$$
$$x = 16$$

2. **The correct answer is (B).** Let the total credit limit be x. Since Mr. Big spends $\dfrac{3}{4}$ of the credit limit, this translates to $\dfrac{3}{4}x$. This is equal to \$9,000, so your equation is $\dfrac{3}{4}x = \$9,000$. Solve for x:

$$\frac{3}{4}x = \$9,000$$
$$3x = 4(\$9,000)$$
$$3x = \$36,000$$
$$x = \$12,000$$

3. **The correct answer is (B).** Let the total number of minutes on the plan be x. Since Brenna has $\dfrac{12}{25}$ of the minutes left, she has used $1 - \dfrac{12}{25} = \dfrac{13}{25}$ of them. This translates to $\dfrac{13}{25}x$. This is equal to 260 minutes, so your equation is $\dfrac{13}{25}x = 260$. Solve for x:

$$\frac{13}{25}x = 260$$
$$13x = 25(260)$$
$$13x = 6,500$$
$$x = 500$$

4. **The correct answer is (E).** Let the total number of minutes of music on the MP3 player be x. Since it is filled to $\frac{4}{5}$ of its capacity, it has $1 - \frac{4}{5} = \frac{1}{5}$ of its capacity left. This translates to $\frac{1}{5}x$. This is equal to 600 minutes, so your equation is $\frac{1}{5}x = 600$. Solve for x:

$$\frac{1}{5}x = 600$$
$$x = 5(600)$$
$$x = 3,000$$

5. **The correct answer is (E).** Henry received four times as many votes as his opponent, so for every 4 that Henry received, his opponent received 1. This means that Henry received $\frac{4}{4+1} = \frac{4}{5}$ of the votes and his opponent received $\frac{1}{4+1} = \frac{1}{5}$ of the votes. Let the total number of votes cast be x. You know that his opponent received 40 votes. Since $\frac{1}{5}$ of votes were for his opponent, this translates to $\frac{1}{5}x = 40$. Solve for x:

$$\frac{1}{5}x = 40$$
$$x = 5(40)$$
$$x = 200$$

chapter 4

Keeping It in Proportion

UNDERSTANDING RATIOS

Ratios are ways to compare things and are really useful in everyday life. You can talk about the ratio of boys to girls at a dance, the ratio of the questions you've answered on a test to all the questions on the test, or the ratio of time you spend in school to the time you spend out of school.

How depressing. I already feel like too much of my life is spent confined within the school walls, but now I'll be able to measure it mathematically.

For example, if there were 10 boys and 8 girls at a dance, you would say that the ratio of boys to girls is 10 to 8. Actually, you'd say that the ratio is 5 to 4 because you always express ratios in their simplest terms. You could also express the ratio of boys to girls as 5:4 or $\frac{5}{4}$. All of these mean the same thing.

The ratio of boys to girls is what we call a *part-to-part* ratio. This means that it compares two or more parts of some bigger whole to each other. There is also a *part-to-whole* ratio that compares one of the parts to the bigger whole of which it is part.

For example, you can write a ratio comparing the number of boys at a dance to all the participants. Since all the participants must be either boys or girls, you know that there are a total of $10 + 8 = 18$ participants. So the ratio of boys to all participants is 10 to 18, which reduces to 5 to 9. It can also be written as 5:9 or $\frac{5}{9}$. The ratio of girls to all participants is 8 to 18, which reduces to 4 to 9. This can also be expressed as 4:9 or $\frac{4}{9}$.

Leave it to the math people. Why have one way to say something when you can make me remember three different ways instead?

Example: Chi spends 9 hours of each day in school. Write this ratio. Next, write the ratio of the amount of time Chi spends in school to the amount of time she spends out of school in a day.

Solution: The first ratio you're asked to write is a part-to-whole ratio. You want to compare the amount of time Chi is in school in a day to the total amount of time in a day. The time spent in school is expressed in hours, but the total time is expressed as a day. Convert the day to hours so that both pieces of the ratio are in the same unit. There are 24 hours in a day, so the amount of time Chi spends in school to the amount of time in a day is 9 to 24, which reduces to 3 to 8. This can also be expressed as 3:8 or $\frac{3}{8}$.

Now you need to write the ratio of time Chi spends in school in a day to the time she doesn't spend in school. This is a part-to-part ratio, since these two amounts of time make up a whole day. You're given the number of hours Chi spends in school but not the number of hours she doesn't. Since the amount of time she's in school, plus the amount of time she's not in school make up the whole day, you can find this missing part. Chi must spend $24 - 9 = 15$ hours out of school. This means that the ratio of time she spends in school to the time she spends out of school in a day is 9 to 15, which reduces to 3 to 5. This could also be written as 3:5 or $\frac{3}{5}$.

That's more information than I need to know about the amount of time I spend in school!

WISE POINTS

★ **Always reduce ratios to simplest terms.**

★ **Always make sure the things you're comparing are in the same unit. If not, convert one so that they are the same unit.**

Get Wise!

Ready for some ratios? Here are a few problems to try.

1. A box contains only chocolate chip and oatmeal cookies. The ratio of chocolate chip cookies to oatmeal cookies in a box is 3:2. What is the ratio of oatmeal cookies to all the cookies in the box?

    (A) 1:3   **(D)** 2:5

    (B) 1:2   **(E)** 3:5

    (C) 2:3

2. A music store orders a total of 300 CDs. If 50 of them are damaged, what is the ratio of damaged to undamaged CDs?

    (A) 1 to 6   **(D)** 5 to 1

    (B) 1 to 5   **(E)** 6 to 1

    (C) 5 to 6

What are they doing, buying CDs that "fell off the back of a truck"?

3. A jar contains *n* nickels, *d* dimes, and *q* quarters. What is the ratio of nickels to total coins in the jar?

    (A) $\dfrac{n}{d+q}$ **(D)** $\dfrac{d}{n+d+q}$

    (B) $\dfrac{d+q}{n}$ **(E)** $\dfrac{n}{n+d+q}$

    (C) $\dfrac{n}{ndq}$

4. There are 5 blue socks, 3 black socks, and 2 white socks in a drawer. What is the ratio of white socks to all the socks in the drawer?

(A) 1:5

(D) 4:1

(B) 1:4

(E) 5:1

(C) 4:5

5. Carrie goes out shopping with $150 in her wallet. When she's done, she has $30 left. What is the ratio of what Carrie spent to what she didn't spend?

(A) 1:4

(D) 4:1

(B) 1:5

(E) 5:4

(C) 4:5

How Wise?

Check your answers on page 93.

USING RATIOS TO FIND ACTUAL NUMBERS

Ratios can also help you solve problems. A ratio gives you information about the relative quantities of items in a situation, which you can use to find actual numbers of items.

> *Example:* A box contains a total of 30 pens, all of which are either blue or black. If the ratio of blue to black pens is 1:2, how many blue pens are in the box?

> *Solution:* Figure out the part-to-whole ratio for blue pens to total pens, and multiply it by the total number of pens.

How boring! Personally, I prefer gel pens.

You know that there are a total of 30 pens. If you can find the ratio of blue pens to all pens, you can multiply it by 30 to find the number of blue pens. You know that the ratio of blue to black pens is 1:2. Since you know that these two parts make up the whole, the whole in this situation can be represented as

$1 + 2 = 3$. This means that the ratio of blue pens to total pens is 1:3 or $\dfrac{1}{3}$. It

makes sense to write the ratio as a fraction here, since you'll be multiplying it by the total to find the actual number:

$$\frac{1}{3} \times 30 = 10$$

So there are a total of 10 blue pens in the box.

WISE NOTE

You can check your answer by figuring out the number of black pens and making sure you get the same ratio the question gave you. Since there are 10 blue pens, there are 30 – 10 = 20 black pens. This means that the ratio of blue to black pens is 10:20, which reduces to 1:2. This is what the question says, so you know that your answer is correct.

Get Wise!

Got it? Then try the following problems where you can use ratios to find actual numbers.

1. The ratio of boys to girls in a math class is 3:4. If there are 28 students in the class, how many of them are girls?

   (A) 4   (D) 16
   (B) 7   (E) 20
   (C) 12

2. There are a total of 12 beverages in a refrigerator, all of which are either sodas or juices. If the ratio of sodas to juices is 3:1, how many of the beverages are sodas?

   (A) 3   (D) 8
   (B) 4   (E) 9
   (C) 6

3. A pet shop sells only cats and dogs. Currently, the store has a total of 60 pets for sale. If the ratio of cats to dogs is 5:7, how many dogs are for sale?

   (A) 12   (D) 45
   (B) 25   (E) 55
   (C) 35

That's a pretty lame pet shop. Dogs and cats are OK, but I'd really love to have a pet monkey!

4. A recipe calls for a total of 6 cups of dry ingredients. Flour and sugar are the only dry ingredients in the recipe. If the ratio of flour to sugar is 7:5, how many cups of sugar are used?

(A) $\dfrac{5}{7}$ (D) $3\dfrac{1}{2}$

(B) $2\dfrac{1}{2}$ (E) 5

(C) $2\dfrac{2}{5}$

5. The Tea Lounge sells a total of 24 different kinds of tea. If the ratio of teas Eileen has tried to those she hasn't is 1:5, how many kinds of tea has Eileen tried?

(A) 4 (D) 18

(B) 6 (E) 20

(C) 8

How Wise?

Check your answers on page 93.

GET INTO PROPORTION

I should be good at these. My mom always says I blow things out of proportion.

When two ratios are equal to each other, they are said to be in proportion or proportional. This fact can be very useful in word problems. If a value in one of the two ratios is unknown, you can find it by setting the ratios equal and then cross-multiplying to solve.

Example: Max is looking at a map. On this map, 3 inches represent an actual distance of 10 feet. If the distance between two towns on the map is 10 inches, what is the actual distance between them?

I'm hopeless at reading maps. The only thing harder than making sense of them is trying to fold them back up.

Solution: Set up a proportion, equating the ratio of the scale of the map to the scale and actual distances between the towns.

You're told that the scale of the map is 3 inches = 10 feet. As a ratio you can write this as $\dfrac{3 \text{ in.}}{10 \text{ ft.}}$. Now write a ratio relating the distance between the two towns on the map and their actual distance. You want to find the actual distance between the two towns, so let that be x. This makes your second ratio $\dfrac{10 \text{ in.}}{x \text{ ft.}}$. Set these two ratios equal and cross-multiply to solve for x:

$$\frac{3 \text{ in.}}{10 \text{ ft.}} = \frac{10 \text{ in.}}{x \text{ ft.}}$$
$$(3)(x) = (10)(10)$$
$$3x = 100$$
$$x = 33\frac{1}{3}$$

So, a distance of 10 inches on the map represents an actual distance of $33\frac{1}{3}$ feet.

WISE NOTE

When setting up a proportion, make sure to put the pieces in the same place in each ratio. In this example, the distance on the map was on the top of the first ratio, so this distance must be on top of the second ratio. If you flip the pieces, you'll have the problem wrong even before you do your calculations. And it's a pretty safe bet that the wrong answer that results from flipping the pieces would be an answer on a multiple-choice question.

Example: If Jacques sells 6 ounces of chocolate for $2.70, how much does he charge for 10 pounds of chocolate?

Solution: Set up a proportion, being sure to make the units the same.

Notice that the first ratio in this question measures chocolate in ounces, while the second does so in pounds. Convert one of these measures before setting up the proportion. It doesn't matter whether you convert ounces to pounds or vice versa, as long as you're consistent. In this case it's easier to convert pounds to ounces (if you converted ounces to pounds you'd have the decimal .375 on the top of your first ratio, which would make your calculations more difficult).

Your first ratio is $\dfrac{6\,\text{oz.}}{\$2.70}$. You're asked for the price of 10 pounds of chocolate, so let that be x. There are 16 ounces in a pound, so there are $10(16) = 160$ ounces in 10 pounds. That makes your second ratio $\dfrac{160\,\text{oz.}}{x}$. Now set the two ratios equal and cross-multiply to solve for x:

$$\frac{6\,\text{oz.}}{\$2.70} = \frac{160\,\text{oz.}}{x}$$
$$6x = 160(\$2.70)$$
$$6x = \$432$$
$$x = \$72$$

$72 on chocolate! I like the stuff but, for that money, I could buy two sweaters, or three DVDs, or five CDs, or . . .

WISE POINT

Units to Keep an Eye On:

* ★ Inches/feet
* ★ Feet/yards
* ★ Centimeters/meters
* ★ Hours/minutes
* ★ Minutes/seconds
* ★ Ounces/pounds

Get Wise!

Now test yourself with the following proportion problems.

1. At a factory, three out of every twenty widgets produced are defective. On Monday, 525 defective widgets are produced. What is the total number of widgets that were produced that day?
    (A) 10,500
    (B) 3,500
    (C) 2,975
    (D) 1,575
    (E) 79

2. Snow is falling at a rate of 2 inches every 25 minutes. At this rate, how long will it take for 4 feet of snow to fall?
    (A) 3.84 minutes
    (B) 50 minutes
    (C) 1 hour and 11 minutes
    (D) 8 hours
    (E) 10 hours

3. Alice draws a floor plan of her bedroom with a scale of 2 inches = 3 feet. If her desk is 4.5 feet long, what will its length be on the floor plan?
    (A) 3 inches
    (B) 3.5 inches
    (C) 4.5 inches
    (D) 5.5 inches
    (E) 6.75 inches

4. If Joan jogs 2.5 miles in 40 minutes, how many miles would she jog in 2 hours?
    (A) .125
    (B) 5
    (C) 7.5
    (D) 15
    (E) 32

5. Robert takes a photograph that has a height of 4 inches and a width of 6 inches. If he enlarges the picture proportionally so that its height is 9 inches, what is its width?

(A) 6 inches   **(D)** 13.5 inches

(B) 7 inches   **(E)** 15 inches

(C) 11 inches

How Wise?

Check your answers on page 95.

A Word to the Wise

* A ratio is a way of comparing things and can be written in three ways:

  1 to 2, 1:2, or $\frac{1}{2}$.

* There are part-to-part ratios and part-to-whole ratios.

* Ratios should be reduced to simplest form and should compare things measured in the same units.

* Multiplying a part-to-whole ratio by the whole amount will give you the actual number for that part.

* When two ratios are equal to each other, they are said to be in proportion.

* You can use a proportion to solve for an unknown value in two equal ratios.

* In a proportion, make sure that your ratios are consistent in order and units.

ANSWERS TO CHAPTER 4: PRACTICE EXERCISES

Understanding Ratios (Page 80)

1. **The correct answer is (D).** Since the box only contains chocolate chip and oatmeal cookies, you can use the part-to-part ratio to write a part-to-whole ratio. The part-to-part ratio is 3:2, so the whole will be represented by $3 + 2 = 5$. That makes the ratio of oatmeal cookies to all cookies 2:5.

2. **The correct answer is (B).** All of the CDs must be either damaged or undamaged. Since 50 of the 300 CDs are damaged, $300 - 50 = 250$ CDs are undamaged. This makes the ratio of damaged to undamaged CDs 50 to 250, which reduces to 1 to 5.

3. **The correct answer is (E).** All the ratios in the answer choices are expressed as fractions, so write your ratio as a fraction. You want the ratio of nickels to total coins. Since there are n nickels, n is the numerator of your fraction. The total coins is the sum of the nickels, dimes, and quarters. Since there are n nickels, d dimes, and q quarters, the total is $n + d + q$. So, the ratio is $\dfrac{n}{n+d+q}$.

4. **The correct answer is (A).** There are 2 white socks in the drawer and a total of $5 + 3 + 2 = 10$ socks. That makes the ratio of white socks to total socks 2:10, which reduces to 1:5.

5. **The correct answer is (D).** Carrie starts with $150 and is left with $30. This means that she spent a total of $150 - $30 = $120. The ratio of what she spent to what she didn't spend is $120:$30, which reduces to 4:1.

Using Ratios to Find Actual Numbers (Page 84)

1. **The correct answer is (D).** The ratio of boys to girls is 3:4. Since all of the students must be either boys or girls, you can use this part-to-part ratio to write a part-to-whole ratio. In this case, the whole will be represented by 3 + 4 = 7. This makes the ratio of girls to total students 4:7. If you write this ratio as a fraction, you can multiply it by the total number of students to find the number of girls:

$$\frac{4}{7} \times 28 = 16$$

2. **The correct answer is (E).** The ratio of sodas to juices is 3:1. Since all of the beverages are either soda or juice, you can use this part-to-part ratio to write a part-to-whole ratio. In this case, the whole will be represented by 3 + 1 = 4. This makes the ratio of sodas to total beverages 3:4. If you write this ratio as a fraction, you can multiply it by the total number of beverages to find the number of sodas:

$$\frac{3}{4} \times 12 = 9$$

3. **The correct answer is (C).** The ratio of cats to dogs is 5:7. Since all of the pets are either cats or dogs, you can use this part-to-part ratio to write a part-to-whole ratio. In this case, the whole will be represented by 5 + 7 = 12. This makes the ratio of dogs to total pets 7:12. If you write this ratio as a fraction, you can multiply it by the total number of pets to find the number of dogs:

$$\frac{7}{12} \times 60 = 35$$

4. **The correct answer is (B).** The ratio of flour to sugar is 7:5. Since all of the dry ingredients are either flour or sugar, you can use this part-to-part ratio to write a part-to-whole ratio. In this case, the whole will be represented by 7 + 5 = 12. This makes the ratio of sugar to total dry ingredients 5:12. If you write this ratio as a fraction, you can multiply it by the total number of cups of dry ingredients to find the number of cups of sugar:

$$\frac{5}{12} \times 6 = \frac{5}{2} = 2\frac{1}{2}$$

5. **The correct answer is (A).** The ratio of teas Eileen has tried to those she has not is 1:5. Since Eileen has either tried a tea or not, you can use this part-to-part ratio to write a part-to-whole ratio. In this case, the whole will be represented by 1 + 5 = 6. This makes the ratio of teas she has tried to total teas 1:6. If you write this ratio as a fraction, you can multiply it by the total number of teas to find the number of teas Eileen has tried:

$$\frac{1}{6} \times 24 = 4$$

Proportion Problems (Page 90)

1. **The correct answer is (B).** Set up a proportion. The first ratio is $\frac{3}{20}$. Since defective widgets are on top in the first ratio, they should also be on top in the second. You're looking for the total number of widgets produced on Monday, so let that be x. This makes your second ratio $\frac{525}{x}$. Set the two ratios equal and cross-multiply to solve for x:

$$\frac{3}{20} = \frac{525}{x}$$
$$3x = (20)(525)$$
$$3x = 10,500$$
$$x = 3,500$$

2. **The correct answer is (E).** Note that one ratio talks about snowfall in inches and the other in feet. Before setting up your proportion, make the units the same. The first ratio is $\frac{2 \text{ in.}}{25 \text{ min.}}$. Since the amount of snowfall is on top in the first ratio, it should be on top in the second as well. Since there are 12 inches in a foot, there are 4(12) = 48 inches in 4 feet. You're looking for the amount of time it will take for 4 feet of snow to fall, so let that be x. This makes your second ratio $\dfrac{48 \text{ in.}}{x \text{ min.}}$. Set the two ratios equal and cross-multiply to solve for x:

$$\frac{2 \text{ in.}}{25 \text{ min.}} = \frac{48 \text{ in.}}{x \text{ min.}}$$
$$2x = (25)(48)$$
$$2x = 1,200$$
$$x = 600$$

So, it will take 600 minutes for the snow to fall. Since there are 60 minutes in an hour, 600 minutes is equal to $\dfrac{600}{60} = 10$ hours.

3. **The correct answer is (A).** Set up a proportion. The first ratio is $\dfrac{2 \text{ in.}}{3 \text{ ft.}}$. Since the number of inches on the floor plan is on top in the first ratio, they should also be on top in the second. You're looking for the number of inches that will represent the desk on the floor plan, so let that be x. This makes your second ratio $\dfrac{x \text{ in.}}{4.5 \text{ ft.}}$. Set the two ratios equal and cross-multiply to solve for x:

$$\frac{2 \text{ in.}}{3 \text{ ft.}} = \frac{x \text{ in.}}{4.5 \text{ ft.}}$$
$$3x = (2)(4.5)$$
$$3x = 9$$
$$x = 3$$

4. **The correct answer is (C).** Note that one ratio talks about time in minutes and the other in hours. Before setting up your proportion, make the units the same. The first ratio is $\dfrac{2.5 \text{ mi}}{40 \text{ min.}}$. Since the number of miles jogged is on top in the first ratio, it should be on top in the second as well. You're looking for the number of miles jogged in 2 hours, so let that be x. Since there are 60 minutes in an hour, there are $2(60) = 120$ minutes in 2 hours. This makes your second ratio $\dfrac{x \text{ mi}}{120 \text{ min.}}$. Set the two ratios equal and cross-multiply to solve for x:

$$\frac{2.5 \text{ mi}}{40 \text{ min.}} = \frac{x \text{ mi}}{120 \text{ min.}}$$
$$40x = (2.5)(120)$$
$$40x = 300$$
$$x = 7.5$$

5. **The correct answer is (D).** Set up a proportion. The first ratio is $\dfrac{4}{6}$. Since height is on top in the first ratio, it should also be on top in the second. You're looking for the width of the enlarged photograph, so let that be x. This makes your second ratio $\dfrac{9}{x}$. Set the two ratios equal and cross-multiply to solve for x:

$$\frac{4}{6} = \frac{9}{x}$$
$$4x = (6)(9)$$
$$4x = 54$$
$$x = 13.5$$

100% Percents

UNDERSTANDING PERCENTS

Percent means "per 100," so a percent can be thought of as a fraction with a denominator of 100. In other words, 15% is the same as $\frac{15}{100}$, which can be reduced to the fraction $\frac{3}{20}$. Percents can also be written as decimals. The decimal equivalent of 15% is .15.

When you're dealing with word problems with percents, you'll find yourself working with the three-piece formula:

$$\text{percent} \times \text{whole} = \text{part}$$

As long as you have two of the three pieces, you can plug them into the formula to solve for the third. The key is to properly identify from the problem which piece is which. Here's how:

★ Percent is the number with the word "percent" or the percent symbol (%) after it.

★ Whole is often referred to as the total and is the number before or after the word "of."

★ "What" or "how many" indicates the unknown value you must find.

So far, this is 100 <u>percent</u> boring. I hope the next <u>part</u> is a <u>whole</u> lot more interesting.

Well, Chi, this is a topic that we'd expect to be near and dear to you. Percents come up in lots of real-life situations, shopping in particular. When you need to figure out how much a $15 CD will cost with 8% sales tax, you use the percent formula. When you need to figure out how much a $40 pair of jeans will cost after a 25% discount, again, it's the percent formula. And if you want to figure out how much of a tip to leave for the generous waitress who gave you extra ice cream with your pie—you got it—it's the percent formula.

Example: Dylan takes a math test that contains a total of 40 questions. If Dylan answers 80% of the questions correctly, how many questions does he answer correctly?

Solution: You're given that the percent is 80% and that the whole (the total number of questions on the test) is 40. Since you have two of the three pieces in the formula, you can plug them in to solve for the third:

$$\text{percent} \times \text{whole} = \text{part}$$
$$80\% \times 40 = \text{part}$$
$$.8 \times 40 = \text{part}$$
$$32 = \text{part}$$

So Dylan answered 32 questions on the test correctly. Note that 80% is converted to the decimal .8 in this solution. You could also have converted it to its fractional equivalent $\frac{4}{5}$ to solve for the part. If you're using a calculator, you can just enter 80 followed by the % sign.

I thought you said that percents had to do with shopping. Here we are in the middle of a math book, talking about a kid taking a math test. Have some pity on me and let's take this lesson to the mall!

OK, OK. Percents do come up a lot in shopping, so let's look at a problem that deals with shopping. The process is the same either way, though, so you'll answer a question about shopping just as you'd answer a question about a math test.

Example: The Music Shack is having a sale of 15% off the regular prices of all CDs. If Ryan saves $9.45 on the CDs he buys, what is the total of the regular prices of the CDs he purchased? (Note: There is no sales tax involved.)

Solution: Identify what you are given. In this case, the percent is 15% and $9.45 is the part. The whole is the total price of the CDs before the sale. Plug the two numbers you are given into the formula and solve for the third:

$$\text{percent} \times \text{whole} = \text{part}$$
$$15\% \times \text{whole} = \$9.45$$
$$.15 \times \text{whole} = \$9.45$$
$$\text{whole} = \frac{\$9.45}{.15} = \$63$$

Get Wise!

Get the hang of it? Then try the following percents problems.

1. Mari treats Joe to lunch at Grimaldi's Pizza. She decides to leave a tip that is equal to 18% of the bill they receive. If Mari leaves a tip of $6.30, what was the total amount of the bill they received?

   (A) $11.34   (D) $42.30
   (B) $24.30   (E) $63.00
   (C) $35.00

> **Pizza—my favorite! The next time Mari feels like treating someone to lunch, tell her to give me a call.**

2. If 18 of the 45 students in the band play woodwind instruments, what percent of the students do NOT play woodwind instruments?

   (A) 2.5%   (D) 60%
   (B) 27%   (E) 97.5%
   (C) 40%

3. Liza needs to get at least 85% of the questions on her final exam correct in order to be on the honor roll. If there are a total of 60 questions on her final exam, at least how many questions must she answer correctly?

   (A) 15   (D) 51
   (B) 25   (E) 85
   (C) 45

4. The population of Springfield increased from 120,000 to 150,000. What percent does this increase represent?

(A) 15% (D) 30%

(B) 20% (E) 35%

(C) 25%

5. Sixty percent of the students in a class are boys. If 12 of the students in the class are girls, what is the total number of students in the class?

(A) 20 (D) 36

(B) 24 (E) 40

(C) 30

A class with 60% boys in it—what's the class and where do I sign up?

How Wise?

Check your answers on page 114.

PERCENT INCREASE OR DECREASE

Many percent word problems deal with a percent increase or decrease. In a percent increase problem, some original whole amount is increased by a certain percent. This results in a new larger whole. When the label on the package of chocolate chip cookies announces that it's now 20% bigger, you know that you'll have more chocolate chip cookies to enjoy. In a percent decrease problem, some original whole amount is decreased by a certain percent. This results in a new smaller whole. If your new and improved toothpaste claims that its users get 8% fewer cavities, it's a good bet you and your dentist will be very happy.

WISE NOTE

Understanding what happens with a percent increase or decrease problem can help you eliminate some answers in multiple-choice questions. In an increase problem, you know that the new whole must be greater than the original whole. That means that you can eliminate choices that are not greater. In a decrease problem, the new whole must be less than the original whole. That means that you can eliminate choices that are not less.

I wish I could decrease the amount of time we're spending talking about percents and increase the amount of time I can spend with my friends!

We're getting there, Chi, in our questions at least. There are a couple of ways to handle percent increase or decrease questions. You can use the three-piece percent formula to solve for the part (the amount of the increase/decrease) and then add or subtract it from the original amount (the whole). You could also add or subtract the percent increase/decrease from 100% (which is the entire whole, the original starting amount) and then multiply the resulting percent by the whole. Use whichever method you feel more comfortable with.

Example: The price of an MP3 player is decreased by 20% for a holiday sale. If the original price of the player is $249, what is the sale price?

Solution: *Method 1. Find part and subtract.* In this question the percent is 20% and the whole is $249. Plug these values into the formula to solve for the part (the amount of the decrease):

$$\text{percent} \times \text{whole} = \text{part}$$
$$20\% \times \$249 = \text{part}$$
$$.2 \times \$249 = \text{part}$$
$$\$49.80 = \text{part}$$

Now that you know that the amount of the decrease is $49.80, you can subtract it from the original price to find the sale price:

$$\$249.00 - \$49.80 = \$199.20$$

Solution: *Method 2. Determine what percent the sale price is of the original price.* In this question you're told that the price of the MP3 player is decreased by 20%. The original price can be thought of as 100%; decreasing it by 20% means that the sale price is 100% − 20% = 80% of the original price. To see what sale price this results in, multiply this percent by the original price:

$$80\% \times \$249 = \text{sale price}$$
$$.8 \times \$249 = \text{sale price}$$
$$\$199.20 = \text{sale price}$$

An MP3 player for less than $200? That's music to my ears! This percent stuff isn't so bad after all. I don't mind figuring out percents if it will help me sniff out bargains like that.

WISE NOTE

One hundred percent of something is the entire thing. In other words, it is equal to 1 times the whole amount in question. When a problem involves less than 100% of something, the resulting amount will be less than the original whole, a fractional part of it, in fact. When a problem involves more than 100% of something, the resulting amount will be more than the original whole, 1 + a fractional part of it.

In other words, 50% of $249 is less than $249, half of it, in fact: $\frac{1}{2} \times \$249 = \$124.50$. Likewise, 200% of $249 is more than $249, 2 times it, in fact: $2 \times \$249 = \$498$.

Example: A retail store sells a jacket for 75% more than the wholesale price it is charged by the manufacturer. If the store sells the jacket for $49, how much is it charged by the manufacturer?

Solution: This one is a little trickier. The question doesn't specifically mention an increase or decrease, so you need to figure out what's going on here. The store is charged a price for the jacket by the manufacturer, which it then increases by 75% to set its retail price. So the retail price is equal to the original price plus 75% of the original price. This means that the retail price is 100% + 75% = 175% of the original price. To see what retail price this results in, multiply this percent by the original price:

$$175\% \times \text{original price} = \$49$$

$$1.75 \times \text{original price} = \$49$$

$$\text{original price} = \frac{\$49}{1.75} = \$28$$

Get Wise!

Time to move on to some percent increase or decrease problems.

1. The price of a sweater after a 10% increase is $44.00. What was the original price of the sweater?

(A) $34.00   (D) $48.40

(B) $39.60   (E) $54.00

(C) $40.00

2. Contributions to a charity totaled $120,000 last year. The contributions decreased by 7% this year. What was the total amount of the contributions made to the charity this year?

(A) $111,600   (D) $127,000

(B) $113,000   (E) $128,400

(C) $117,000

3. The average attendance at a baseball game this year was 35,000 people. This was an increase of 12% from last year. What was the average attendance at a baseball game last year?

(A) 47,000   (D) 31,250

(B) 39,200   (E) 23,000

(C) 33,000

And I'll bet I know why attendance is up this year. Have you seen the team's new shortstop? He's 100% cuter than the old one.

4. Jorge has a coupon for 10% off the price of a meal at Hamburger Hut. If he pays $3.60 for his meal using the coupon, what was the original price of the meal? (Note: There is no sales tax.)

(A) $3.24

(D) $3.96

(B) $3.50

(E) $4.00

(C) $3.70

5. Jane receives a salary increase of 5% each year. If her current annual salary is $40,000, what will her annual salary be after *two* years?

(A) $42,000

(D) $45,000

(B) $44,000

(E) $50,000

(C) $44,100

How Wise?

Check your answers on page 116.

Get Wise! Mastering Math Word Problems

PROBLEMS WITHOUT A WHOLE

Some of the most difficult percent questions involve situations where you aren't given a whole. You're told about a percent change or changes and are asked what effect they'll have on the whole. So how do you solve a problem where you aren't given a whole? You make one up.

Make one up? What is this, the land of make-believe? Are we writing fiction here? I didn't think that you could do creative stuff like that in math.

It's perfectly all right to make up a number for the whole. You use this number to apply the percent change(s), and see what result you get. If the question has multiple choices, use this same number in each answer, and select the one that matches your result. It doesn't matter what number you use. As long as you use the same number throughout the problem, your results will be correct.

Since it doesn't matter mathematically what number you use, it makes sense to use the number that's easiest to work with. For percent problems, that number is generally 100 because it's so easy to take percents of 100: 10% of 100 is 10, 17% of 100 is 17, 45.678% of 100 is 45.678, and so on.

Example: Increasing a number by 20% and then increasing this result by 10% is the same as increasing the original number by what percent?

Solution: A common mistake to make on this question is to think that an increase of 20% followed by an increase of 10% is the same as a 30% increase. This isn't the case because you can only add and subtract percents of the same whole. In this question, the 20% increase is applied to the original whole, but the 10% increase is applied to the larger whole that results from the first increase.

Solve this by making up a number for the whole and seeing what happens. Use 100 since this is a percent problem.

Since 20% of 100 is 20, increasing 100 by 20% results in the new whole 100 + 20 = 120. Since 10% of 120 is 12, increasing 120 by 10% results in the new whole 120 + 12 = 132. Since the original whole was 100, this represents an increase of 132 − 100 = 32. Since 32 is 32% of 100, this is the same as increasing the original number by 32%.

Get Wise!

Now test yourself with some percents problems without wholes.

1. The price of a stereo is reduced by 20% for a sale. By what percent must the sale price be increased to return the stereo to its original price?

(A) 10%   (D) 25%

(B) 20%   (E) 30%

(C) 22.5%

2. The price of a share of stock increases by 20% on Monday. On Tuesday, the price of the share of stock increases by a further 15%. This is equivalent to the original price of the stock being increased by which percent?

(A) 5%   (D) 35%

(B) 17.5%   (E) 38%

(C) 30%

3. A number is increased by 10% and the result is then decreased by 25%. This is the same as decreasing the original number by what percent?

(A) 12.5%   (D) 35%

(B) 15%   (E) 82.5%

(C) 17.5%

4. At a certain store, the price of a handbag is reduced by 25% for a sale. A week later this sale price is reduced by an additional 25% for an end-of-season clearance. What percent of the original price is the clearance price?

(A) 42.5%   (D) 66%

(B) 50%   (E) 75%

(C) 56.25%

That's a major sale. When we finish up with this chapter, I'm headed straight to a certain store to put my percents education to practical use.

5. Decreasing a number *n* by 10% and then increasing the result by 30% is equivalent to

(A) decreasing *n* by 10 percent.

(B) *n*.

(C) increasing *n* by 10 percent.

(D) increasing *n* by 17 percent.

(E) increasing *n* by 20 percent.

How Wise?

Check your answers on page 118.

A Word to the Wise

Percents Formula:

- ★ The formula to use for percents word problems is **percent × whole = part.** If you know two of the three pieces of this formula, you can plug in to find the third.

- ★ When identifying the three pieces of the percent formula from a word problem, remember: The percent is the number followed by the percent sign or the word "percent"; the whole is often indicated by the word "total" and is the number before or after the word "of"; the unknown quantity that you must solve for is often indicated by the words "what" or "how many."

Percent Increase or Decrease:

- ★ To find the result of a percent increase, use the three-piece formula to solve for the part and add it to the original whole.

- ★ To find the result of a percent decrease, use the three-piece formula to solve for the part and subtract it from the original whole.

- ★ Alternatively, you can figure out what percent of the whole this change will result in, and then multiply it by the original whole.

Percents Problems Without Wholes:

- ★ If you aren't given a whole in a percent problem, use 100 as the whole. Apply the percent changes to it and see what happens.

ANSWERS TO CHAPTER 5: PRACTICE EXERCISES

Percents Problems (Page 102)

1. **The correct answer is (C).** Use the three-piece formula. The percent is 18% and the part is $6.30. Plug in and solve for the whole:

$$\text{percent} \times \text{whole} = \text{part}$$
$$18\% \times \text{whole} = \$6.30$$
$$.18 \times \text{whole} = \$6.30$$
$$\text{whole} = \frac{\$6.30}{.18} = \$35$$

2. **The correct answer is (D).** Use the three-piece formula. Be careful identifying the part. You're asked to find the percent of students who do NOT play woodwind instruments. Since 18 students play them, 45 − 18 = 27 students do not play woodwind instruments. So, the whole is 45 and the part is 27. Plug in and solve for the percent:

$$\text{percent} \times \text{whole} = \text{part}$$
$$\text{percent} \times 45 = 27$$
$$\text{percent} = \frac{27}{45} = \frac{3}{5}$$
$$\text{percent} = 60\%$$

3. **The correct answer is (D).** Use the three-piece formula. The percent is 85% and the whole is 60. Plug in and solve for the whole:

$$\text{percent} \times \text{whole} = \text{part}$$
$$85\% \times 60 = \text{part}$$
$$.85 \times 60 = \text{part}$$
$$51 = \text{part}$$

4.   **The correct answer is (C).** Use the three-piece formula. You're asked to find the percent, so the question must provide you with the part and the whole. You need to do some work to identify them though. The population starts at 120,000 and jumps to 150,000, which is an increase of 150,000 – 120,000 = 30,000. This is the part. The whole is the original amount. Since the population started at 120,000, this is the whole. (150,000 is the new whole you get from adding the part to the original whole.) Plug in and solve for the percent:

$$\text{percent} \times \text{whole} = \text{part}$$
$$\text{percent} \times 120,000 = 30,000$$
$$\text{percent} = \frac{30,000}{120,000} = \frac{1}{4}$$
$$\text{percent} = 25\%$$

5.   **The correct answer is (C).** Use the three-piece formula. Be careful identifying the percent. The question says that 12 of the students in the class are girls, so that is the part. Since 60% of the students are boys, 100% – 60% = 40% must be girls. Plug in and solve for the whole:

$$\text{percent} \times \text{whole} = \text{part}$$
$$40\% \times \text{whole} = 12$$
$$.4 \times \text{whole} = 12$$
$$\text{whole} = \frac{12}{.4} = 30$$

Percent Increase or Decrease (Page 107)

1. **The correct answer is (C).** First, think about the size of the answer. Since the price after an increase is $44, the original price must be less than $44. This allows you to eliminate answer choices (D) and (E). Since the price of $44 represents a 10% increase of the original price, it represents 10% + 100% = 110% of the original price. So this percent multiplied by the original price should result in $44:

$$110\% \times \text{original price} = \$44$$
$$1.1 \times \text{original price} = \$44$$
$$\text{original price} = \frac{\$44}{1.1} = \$40$$

2. **The correct answer is (A).** First, think about the size of the answer. The contributions have decreased since last year, so this year's contributions must be less than $120,000. This allows you to eliminate answer choices (D) and (E). Since this year's contributions are 7% less than last year's, they represent 100% – 7% of last year's total. So this percent multiplied by last year's total will give you this year's total:

$$93\% \times \$120,000 = \text{this year's total}$$
$$.93 \times \$120,000 = \text{this year's total}$$
$$\$111,600 = \text{this year's total}$$

Alternatively, you could have used the three-part formula to calculate the part (the amount of decrease in contributions) and then subtracted that from last year's total:

$$\text{percent} \times \text{whole} = \text{part}$$
$$7\% \times \$120,000 = \text{part}$$
$$.07 \times \$120,000 = \text{part}$$
$$\$8,400 = \text{part}$$
$$\text{last year's total} - \text{amount of decrease} = \text{this year's total}$$
$$\$120,000 - \$8,400 = \text{this year's total}$$
$$\$111,600 = \text{this year's total}$$

3. **The correct answer is (D).** First, think about the size of the answer. Since the attendance after an increase is 35,000, the original attendance must be less than 35,000. This allows you to eliminate answer choices (A) and (B). Since the attendance of 35,000 represents a 12% increase from last year, it represents 12% + 100% = 112% of last year's attendance. So this percent multiplied by last year's attendance should result in 35,000:

$$112\% \times \text{last year's attendance} = 35,000$$
$$1.12 \times \text{last year's attendance} = 35,000$$
$$\text{last year's attendance} = \frac{35,000}{1.12} = 31,250$$

4. **The correct answer is (E).** First, think about the size of the answer. Since the price of the meal after a decrease is $3.60, the original price must be greater than that. This allows you to eliminate answer choices (A) and (B). Since the price of $3.60 represents a 10% decrease from the original price, it represents 100% – 10% = 90% of the original price. So this percent multiplied by the original price should result in $3.60:

$$90\% \times \text{original price} = \$3.60$$
$$.9 \times \text{original price} = \$3.60$$
$$\text{original price} = \frac{\$3.60}{.9} = \$4.00$$

5. **The correct answer is (C).** First, think about the size of the answer. Since Jane's salary will be increased by 5% each year for two years, the new salary must be greater than $40,000. All of the choices are greater, though, so you can't eliminate any choices. A common mistake is to assume that two annual increases of 5% is equal to one annual increase of 10%. This isn't true, since after the first year the 5% raise is calculated on a higher annual salary.

Jane's salary after one year will be 5% + 100% = 105% of her current salary:

$$105\% \times \$40,000 = \text{salary after one year}$$
$$1.05 \times \$40,000 = \text{salary after one year}$$
$$\$42,000 = \text{salary after one year}$$

Her salary after two years will be 5% + 100% = 105% of this new salary:

$$105\% \times \$42,000 = \text{salary after two years}$$
$$1.05 \times \$42,000 = \text{salary after two years}$$
$$\$44,100 = \text{salary after two years}$$

Percents Problems Without a Whole (Page 111)

1. **The correct answer is (D).** Remember, you can't add and subtract percents of different wholes. Make up a number for the price of the stereo. Use $100, since it's easy to find percents of 100. Since 20% of $100 is $20, reducing the price by 20% results in a price of $80. In order for the sale price of the stereo to return to its original price, it must increase by $20. The sale price is $80 and the part is $20, so plug into the three-piece formula to see what percent this represents:

$$\text{percent} \times \text{whole} = \text{part}$$
$$\text{percent} \times \$80 = \$20$$
$$\text{percent} = \frac{\$20}{\$80} = \frac{1}{4}$$
$$\text{percent} = 25\%$$

2. **The correct answer is (E).** Remember, you can't add and subtract percents of different wholes. Make up a number for the price of the stock. Use $100, since it's easy to find percents of 100. Since 20% of $100 is $20, increasing the price by 20% results in a price of $120. Since 15% of $120 is $18, increasing this price by 15% results in a price of $138. Since the original price of the stock was $100, this represents an increase of $38. Since $38 is 38% of $100, this is an overall increase of 38% of the original price of the stock.

3. **The correct answer is (C).** Remember, you can't add and subtract percents of different wholes. Make up a number for the whole. Use 100, since it's easy to find percents of 100. Since 10% of 100 is 10, increasing the number by 10% results in 110. Now you have to decrease 110 by 25%, which is 27.5. 110 − 27.5 = 82.5. Since the original number was 100, this represents a decrease of 17.5. Since 17.5 is 17.5% of 100, this is an overall decrease of 17.5% of the original number.

4. **The correct answer is (C).** Remember, you can't add and subtract percents of different wholes. Make up a number for the price of the handbag. Use $100, since it's easy to find percents of 100. Since 25% of $100 is $25, reducing the price by 25% results in a price of $75. Since 25% of $75 is $18.75, reducing this sale price by 25% results in a clearance price of $56.25. Since the original price was $100, the clearance price represents 56.25% of the original price.

5. **The correct answer is (D).** Remember, you can't add and subtract percents of different wholes. Make up a number for the whole n. Use 100, since it's easy to find percents of 100. Since 10% of 100 is 10, decreasing the number by 10% results in 90. Since 30% of 90 is 27, increasing this result by 30% results in 117. Since the original number was 100, this represents an increase of 17. Since 17 is 17% of 100, this is an overall increase of 17% of the original number.

chapter 6

Not Just Your Average Chapter

UNDERSTANDING AVERAGES

Averages? Who wants to focus on being run of the mill and ordinary?

Averages are technically called "measures of central tendency" and are used to represent information about sets of data. In the world of statistics, a "measure of central tendency" is one number that quickly indicates what all the numbers in the set represent as a whole. This number basically represents the average of all the numbers in the set. You've probably heard politicians spouting statistics in their speeches, or seen journalists using them in articles in the newspaper. If you have a sense of what the different measures mean, you'll be better able to understand what the people using them are trying to say.

I have a tendency to doze off when people drone on about statistics.

When most people think of averages, they think of the mean. The mean is often referred to as just plain old average in everyday life. For example, the mean is what your teacher is talking about when she says that the average grade on the test was 81. Another example of mean is what a baseball card refers to when it lists a player's batting average. So, in other words, the mean is the average that you'll have to deal with most often. We'll look at its formula and focus on it for most of this chapter. First, let's get the two other types of averages out of the way: median and mode.

Wait, don't tell me. A median is that thing in the middle of the road, that's there so you don't drive into oncoming traffic. And mode . . . well, I think mode has something to do with ice cream. You know, pie à la mode.

Almost, Chi. Your guess about median isn't too far off. The median is the middle value in a set of data that is organized numerically. In other words, it is the middle number in a set of numbers arranged from smallest to largest or largest to smallest. If there is an even number of values in the set, the median is the sum of the middle two values divided by 2. (This is actually the mean of the middle two values, but we'll get to that later.)

The mode has nothing to do with ice cream, gelato, sorbet, or any other kind of frozen dessert. The mode is the value that occurs most often in a set of data. If all the values occur the same number of times, the set has no mode. If there is a tie between two or more terms for which occurs most often, the set has more than one mode.

Example: Jackson buys five DVDs. Their prices are as follows: $15, $18, $12, $14, $20. What is the median price of a DVD that Jackson buys?

Solution: The first step to finding the median is to put the numbers in order. It's easiest to do this from smallest to largest, so do that: $12, $14, $15, $18, $20.

Once the numbers are in order, all you have to do is pick the middle value. Since there are five numbers in the set, the middle value is the third: $15.

Suppose that Jackson thought that $20 was too steep, and only bought the four cheaper DVDs. That would make the values in order $12, $14, $15, $18. Since there is an even number of terms, there is no one middle value. That means that the median is the sum of the two middle values, divided by 2:

$$\frac{\$14 + \$15}{2} = \frac{\$29}{2} = \$14.50.$$

Jackson should join one of those DVD clubs. You can get ten DVDs for just a penny—how's that for a median price?

Example: The coach recorded the number of points scored by the basketball team in each of its last seven games, as shown in the following chart. What is the mode number of points scored by the basketball team for these games?

Game 1	Game 2	Game 3	Game 4	Game 5	Game 6	Game 7
69	63	87	72	69	78	87

The mode is the value that appears most often. Check the chart to see which, if any, values turn up most often. The value 69 appears twice, and so does the value 87. The rest of the values appear only once. This means that the set has two modes: 69 and 87.

WISE POINTS

★ A set of numbers will always have a median. The median will either be a number in that set, or the mean of two numbers in the set.

★ A set of numbers may or may not have a mode. If a set does have a mode (or modes), the mode will be a number in the set.

Get Wise! Mastering Math Word Problems

Get Wise!

Here are some problems to help you understand averages.

1. Kecia baby-sat for the neighbors five times in a one-month period. The amounts of money she earned were $16.00, $24.50, $20.00, $18.50, and $16.00. What was the median amount of money that Kecia earned?

(A) $16.00   (D) $20.00

(B) $18.50   (E) $24.50

(C) $19.00

2. Dave used the Internet to research the average monthly temperatures on a Caribbean island. The information he found is printed below. What is the mode monthly temperature for this island?

March	April	May	June	July	August
78°	83°	87°	81°	85°	87°

(A) 78°   (D) 85°

(B) 81°   (E) 87°

(C) 83°

3. Mrs. Crabapple gave her class an algebra test. The scores for this test were: 94, 73, 82, 90, 89, 82, 73, 89, 86, and 82. What was the mode score for this test?

(A) 73   (D) 89

(B) 82   (E) 94

(C) 84

4. What was the median score for Mrs. Crabapple's algebra test?

(A) 73   (D) 89

(B) 82   (E) 94

(C) 84

5. The prices for five candy treats at a movie theater are as follows: $1.75, $2.00, $1.25, $1.50, $2.00. What is the median price of a candy treat?

(A) $1.25

(D) $1.75

(B) $1.50

(E) $2.00

(C) $1.60

How Wise?

Check your answers on page 141.

AVERAGE (ARITHMETIC MEAN) FORMULA

Like we said, the average you'll see most often is the mean. If a word problem talks about an average and doesn't specify mean, median, or mode, it's talking about the mean.

If someone had the nerve to call me average, I'd think that they were really mean.

Sometimes, a question will use the word "average," but then in parentheses tell you it's the arithmetic mean you're looking for. You can use a three-piece formula to find this average:

$$\text{Average} = \frac{\text{Sum of terms}}{\text{Number of terms}}$$

So, you add up the group of numbers you're given—whether they're grades, prices, or temperatures—and then you divide by the number of numbers in the group. Simple, huh?

So, I'll use the average formula for the average word problem. I'm starting to get dizzy . . .

Example: Pat bought the first five Harry Potter books for the following prices: $17.95, $17.95, $19.95, $25.95, and $29.95. What was the average price of one of these books?

Solution: Use the average formula. All you have to do is add up the prices of all the books, and then divide by 5:

$$\text{Average} = \frac{\text{Sum of terms}}{\text{Number of terms}}$$

$$\text{Average} = \frac{\$17.95 + \$17.95 + 19.95 + \$25.95 + \$29.95}{5}$$

$$\text{Average} = \frac{\$111.75}{5}$$

$$\text{Average} = \$22.35$$

Although it's the most expensive, I'd say the last book is a bargain. The first book costs $17.95 but it only has 309 pages, so the average price of a page is $\frac{\$17.95}{309} \approx \$0.06$. The last book costs $29.95 but it has 869 pages, so its average price per page is $\frac{\$29.95}{809} \approx \$0.04$. I think I've been spending too much time with this book—I'm starting to sound like a real math geek!

Since this is a three-piece formula, as long as you have any two pieces you can always solve for the third.

Example: Johnny's batting average for this season was .340. If he had 600 at bats, how many base hits did he have? (Batting average is equal to number of base hits divided by total number of at-bats.)

Solution: Use the average formula. In this case you're given the average and the number of terms. Plug these numbers into the formula to solve for the sum of the terms (the number of base hits):

$$\text{Average} = \frac{\text{Sum of terms}}{\text{Number of terms}}$$

$$.340 = \frac{\text{Sum of terms}}{600}$$

$$.340 \times 600 = \text{Sum of terms}$$

$$204 = \text{Sum of terms}$$

Get Wise!

Ready to test your skill at using the average formula? Then check out these problems.

1. Michael's grades on his math tests are as follows: 84, 79, 90, 87, and 90. What is Michael's average grade for these tests?

   (A) 79   **(D)** 87

   (B) 84   **(E)** 90

   (C) 86

2. Eileen went on a shopping spree at the mall. She visited eight stores and spent an average of $40 at each. What is the total amount of money that Eileen spent?

   (A) $5   **(D)** $320

   (B) $48   **(E)** $408

   (C) $140

3. The total number of people who went to the Cineplex on Friday to see the new horror movie was 1,305. If an average of 145 people attended each screening, how many screenings were there?

   (A) 9   **(D)** 1,450

   (B) 45   **(E)** 189,225

   (C) 1,160

You want horror? Think about the average price of a ticket, popcorn, and a drink at the movies these days. Now that's scary!

4. John has five cousins whose ages are 8, 11, 12, 5, and 9. What is the average age of John's cousins?

(A) 5   (D) 11

(B) 8   (E) 12

(C) 9

5. The following chart shows how many hours per day Ally worked at her after-school job. What is the average number of hours that Ally worked for these five days?

Monday	Tuesday	Wednesday	Thursday	Friday
2	0	3	3	5

(A) 2   (D) 3.25

(B) 2.6   (E) 5

(C) 3

How Wise?

Check your answers on page 141.

WORKING WITH THE SUM

Say that you have a so-so average on your three geometry tests, and there's one more test coming up before you get your final grade for the semester. Your parents made it clear that they'd like you to bring your average up four points by the end of the semester. Is it possible? If so, what score do you need to get on your last test? And most important, how on earth do you figure it out?

I've made it clear to my parents that I'd like a bigger allowance and a later curfew, but I haven't had any luck so far. I wonder if there's a test score high enough to convince them to do that.

Example: Your average grade for your first three tests is 81. If there is one test left and you want to raise your average to 85, what grade do you need to get on the final test?

Solution: Find the sum of what you want, and the sum of what you already have. The difference between them will help you figure out what you need.

You need to use the three-piece average formula for this question. But do you have two pieces? The question says that your average now is 81 and that you want it to be 85. That's two numbers for the average and none for the sum or number of terms.

Let's think about it a minute. You've taken three tests, and there's one more left. You want your average for all four of these tests to be 85. That means that the average is 85 and the number of terms is 4. Plug into the formula and solve for the sum:

$$\text{Average} = \frac{\text{Sum of terms}}{\text{Number of terms}}$$

$$85 = \frac{\text{Sum of terms}}{4}$$
$$85 \times 4 = \text{Sum of terms}$$
$$340 = \text{Sum of terms}$$

This means that for your average for all four tests to be 85, the total of all four scores must be 340. So how does this help you figure out what score you need on the last test? Well, you know that the average for your first three tests was 81. If you plug these values into the formula, you can solve for the sum of your first three test grades:

$$81 = \frac{\text{Sum of terms}}{3}$$
$$81 \times 3 = \text{Sum of terms}$$
$$243 = \text{Sum of terms}$$

Since you need 340 points and you already have 243, you need $340 - 243 = 97$ points on your last test. Better get studying now!

A 97! That's some score. If I got a 97, I'd definitely expect a bigger allowance, a later curfew, and a month where my little brother does all my chores.

Get Wise!

Now try the following problems where you'll have to work with the sum.

1. Ty bowls three games. His scores for his first two games are 119 and 133. If he wants to finish with an average score of 125 for all three games, what score must he bowl on his last game?

(A) 14   (D) 126

(B) 123   (E) 375

(C) 125

2. Lori spent an average of 3 hours each weeknight studying last week. If she spent an average of 3.5 hours each night from Monday through Thursday, how many hours did she spend studying Friday night?

(A) 1   (D) 12

(B) 3.5   (E) 15

(C) 7

Studying on a Friday night? What kind of a book is this?

3. The high school is holding a fund-raiser. The ninth, tenth, and eleventh grades have raised an average of $425 each. If the goal of the fund-raiser is for each grade to raise an average of $500, how much money must the twelfth-grade class raise to meet this goal?

(A) $75   (D) $725

(B) $225   (E) $925

(C) $462

4. The average of eight numbers is 11. If the average of seven of the numbers is 12, what is the eighth number?

(A) 4   (D) 84

(B) 11.5   (E) 88

(C) 56

5. The average of five numbers is 24. If the number 30 is removed from the set, what is the average of the remaining four numbers?

(A) 18   (D) 90

(B) 22.5   (E) 120

(C) 27

How Wise?

Check your answers on page 143.

WEIGHTED AVERAGES

What if you want to find the average of two averages? Say, for example, that for a particular test, you know the average grade of the boys in your class and the average grade of the girls. How would you find the average for the whole class?

Hey, as long as I do well on a test, I'm not too concerned about everyone else's grades.

Example: There are eight boys and twelve girls in the class. If the boys' average score for the test is 78 and the girls' average score is 85, what is the average grade for the entire class?

Solution: "Weight" each average according to the number of people or things it represents.

The common mistake is to simply average the two averages, and decide that the class average is $\dfrac{78+85}{2} = \dfrac{163}{2} = 81.5$. The problem is that you can't average two averages that represent different numbers of terms. Since the 78 average represents eight boys and the 85 represents twelve girls, you can't just average them. The 85 must get more "weight" in the average since it represents more students' scores.

$$\text{Average} = \frac{\text{Sum of terms}}{\text{Number of terms}}$$

$$\text{Average} = \frac{8(78)+12(85)}{8+12}$$

$$\text{Average} = \frac{8(78)+12(85)}{20}$$

$$\text{Average} = \frac{\overset{2}{\cancel{8}}(78) + \overset{3}{\cancel{12}}(85)}{\underset{5}{\cancel{20}}}$$

$$\text{Average} = \frac{156 + 255}{5}$$

$$\text{Average} = \frac{411}{5}$$

$$\text{Average} = 82.2$$

WISE NOTE

To see if you're dealing with a weighted average, check to see how many people or things each average represents. If the averages represent different numbers of items, it's a weighted average. If the averages represent the same number of items, it's OK to just average the averages.

You know something about the total average when you're dealing with a weighted average. Since the 85 average represents more terms than the 78 average, the total average must be closer to 85 than 78.

Get Wise!

Ready to work on some problems with weighted averages? Then try these.

1. Jamie plays a video game three times before dinner, and his average score is 987. He plays the same game two times after dinner and his average score for these games is 1,202. What is Jamie's average score for all five games?

   (A) 1,073   (D) 1,201

   (B) 1,094.5   (E) 5,365

   (C) 1,116

2. The average of three numbers is 15. The average of five other numbers is 7. What is the average of all eight of these numbers?

   (A) 8   (D) 12

   (B) 10   (E) 22

   (C) 11

3. Hermione's average grade on her first five history tests is 93. If she gets a grade of 87 on her sixth and final test, what is her average history grade?

   (A) 88   (D) 94

   (B) 90   (E) 96

   (C) 92

4. Greg went holiday shopping at the mall. He spent an average of $22 on the gifts he bought for his two sisters. He spent an average of $15 on the gifts that he bought for his three cousins. What is the average amount of money that Greg spent on these five gifts?

   (A) $7.00   (D) $19.20

   (B) $17.80   (E) $37.00

   (C) $18.50

Greg sounds pretty generous. Maybe he could talk to my little brother about the joy of giving. For my last birthday, all he gave me was a pain in the neck.

5. During the five days of the week, soccer practice lasts an average of 1.6 hours. On the two weekend days, soccer practice lasts an average of 3 hours. What is the average length of a soccer practice, in hours, for the entire week?

(A) 2

(B) 2.3

(C) 2.6

(D) 4.6

(E) 14

How Wise?

Check your answers on page 144.

A Word to the Wise

★ The median is the middle number in a group of numbers arranged in numerical order.

★ The mode is the value that occurs most often in a group of numbers.

★ The mean is often referred to as the average and is found using the three-piece formula:

$$\text{Average} = \frac{\text{Sum of terms}}{\text{Number of terms}}$$

★ If a question gives you one average and asks what you need to result in another average, the key is to work with the sum.

★ You can't average averages if they represent different numbers of items. "Weight" each average according to the number of items it represents.

ANSWERS TO CHAPTER 6: PRACTICE EXERCISES

Understanding Averages (Page 125)

1. **The correct answer is (B).** Arrange the amounts in order from smallest to largest: $16.00, $16.00, $18.50, $20.00, $24.50. Since there is an odd number of terms in this group, the median is the middle value: $18.50.

2. **The correct answer is (E).** The mode is the value that occurs most often in a group of numbers. In this group, 87° appears twice, while all the other values appear only once. So the mode is 87°.

3. **The correct answer is (B).** The mode is the value that occurs most often in a group of numbers. In this group, several values appear more than once: 73 appears twice, as does 89. However, the value 82 appears three times, so it is the mode.

4. **The correct answer is (C).** Arrange the scores in order from smallest to largest: 73, 73, 82, 82, 82, 86, 89, 89, 90, 94. Since there is an even number of terms in the group, the median is the mean of the two middle values:

$$\text{median} = \frac{82 + 86}{2} = \frac{168}{2} = 84$$

5. **The correct answer is (D).** Arrange the prices in order from smallest to largest: $1.25, $1.50, $1.75, $2.00, $2.00. Since there is an odd number of terms in the group, the median is the middle value: $1.75.

Average Formula (Page 130)

1. **The correct answer is (C).** You're given five test scores, so the number of terms is 5. The sum of the terms is the sum of test scores. Plug these values into the average formula to solve for the average:

$$\text{Average} = \frac{\text{Sum of terms}}{\text{Number of terms}} = \frac{84 + 79 + 90 + 87 + 90}{5} = \frac{430}{5} = 86$$

2. **The correct answer is (D).** You're given that the average is $40 and
    the number of terms is 8. Plug these values into the average formula
    to solve for the sum of terms:

    $$\text{Average} = \frac{\text{Sum of terms}}{\text{Number of terms}}$$

    $$\$40 = \frac{\text{Sum of terms}}{8}$$

    $$\$40 \times 8 = \text{Sum of terms}$$
    $$\$320 = \text{Sum of terms}$$

3. **The correct answer is (A).** You're given that the average is 145 and
    the sum of terms is 1,305. Plug these values into the average formula
    to solve for the number of terms:

    $$\text{Average} = \frac{\text{Sum of terms}}{\text{Number of terms}}$$

    $$145 = \frac{1,305}{\text{Number of terms}}$$

    $$145 \times \text{Number of terms} = 1,305$$

    $$\text{Number of terms} = \frac{1,305}{145} = 9$$

4. **The correct answer is (C).** You're given five ages, so the number of terms is 5. The sum of the terms is the sum of ages. Plug these values into the average formula to solve for the average:

$$\text{Average} = \frac{\text{Sum of terms}}{\text{Number of terms}} = \frac{8+11+12+5+9}{5} = \frac{45}{5} = 9$$

5. **The correct answer is (B).** You're given information about the hours Ally worked on five days, so the number of terms is 5. The sum of the terms is the sum of the hours. Plug these values into the average formula to solve for the average:

$$\text{Average} = \frac{\text{Sum of terms}}{\text{Number of terms}} = \frac{2+0+3+3+5}{5} = \frac{13}{5} = 2.6$$

Working with the Sum (Page 134)

1. **The correct answer is (B).** In order to have an average of 125 for all three games, Ty must score a total of $125 \times 3 = 375$ points. Since he already has a total of 119 + 133 = 252 points, Ty must score 375 – 252 = 123 points.

2. **The correct answer is (A).** In order to have spent an average of 3 hours for each of five days studying, Lori must have spent a total of 3 × 5 = 15 hours studying. Since she spent 3.5 × 4 = 14 hours studying from Monday through Thursday, she must have spent 15 – 14 = 1 hour studying Friday.

3. **The correct answer is (D).** In order for the four grades in the high school to raise an average of $500 each, they must raise a total of $500 × 4 = $2,000. Since three grades have already raised $425 × 3 = $1,275, the twelfth grade must raise $2,000 – $1,275 = $725.

4. **The correct answer is (A).** If the average of eight numbers is 11, their sum is 11 × 8 = 88. If the average of seven of these numbers is 12, then the sum of these seven numbers is 12 × 7 = 84. That means that the remaining number must be 88 – 84 = 4.

5. **The correct answer is (B).** If the average of five numbers is 24, then their sum is $24 \times 5 = 120$. If the number 30 is removed from this set of numbers, the sum of the remaining four must be $120 - 30 = 90$.

Since the sum of the 4 numbers is 90, their average must be $\dfrac{90}{4} = 22.5$.

Weighted Averages (Page 138)

1. **The correct answer is (A).** The average of 987 represents three scores, and the average of 1,202 represents two scores. Plug into the average formula, weighting each average accordingly:

$$\text{Average} = \frac{\text{Sum of terms}}{\text{Number of terms}}$$

$$\text{Average} = \frac{3(987) + 2(1,202)}{5}$$

$$\text{Average} = \frac{2,961 + 2,404}{5}$$

$$\text{Average} = \frac{5,365}{5} = 1,073$$

2. **The correct answer is (B).** The average of 15 represents three numbers and the average of 7 represents five numbers. Plug into the average formula, weighting each average accordingly:

$$\text{Average} = \frac{\text{Sum of terms}}{\text{Number of terms}}$$

$$\text{Average} = \frac{3(15) + 5(7)}{8}$$

$$\text{Average} = \frac{45 + 35}{8}$$

$$\text{Average} = \frac{80}{8} = 10$$

3. **The correct answer is (C).** The average of 93 represents five test scores and the score of 87 represents one score. Plug into the average formula, weighting each accordingly:

$$\text{Average} = \frac{\text{Sum of terms}}{\text{Number of terms}}$$

$$\text{Average} = \frac{5(93) + 1(87)}{6}$$

$$\text{Average} = \frac{465 + 87}{6}$$

$$\text{Average} = \frac{552}{6} = 92$$

4. **The correct answer is (B).** The average of $22 represents two gifts and the average of $15 represents three gifts. Plug into the average formula, weighting each average accordingly:

$$\text{Average} = \frac{\text{Sum of terms}}{\text{Number of terms}}$$

$$\text{Average} = \frac{2(\$22) + 3(\$15)}{5}$$

$$\text{Average} = \frac{\$44 + \$45}{5}$$

$$\text{Average} = \frac{\$89}{5} = \$17.80$$

5. **The correct answer is (A).** The average of 1.6 represents five days and the average of 3 represents two days. Plug into the average formula, weighting each average accordingly:

$$\text{Average} = \frac{\text{Sum of terms}}{\text{Number of terms}}$$

$$\text{Average} = \frac{5(1.6) + 2(3)}{7}$$

$$\text{Average} = \frac{8 + 6}{7}$$

$$\text{Average} = \frac{14}{7} = 2$$

chapter 7

Math in Motion

UNDERSTANDING MOTION PROBLEMS

One classic word problem type is the motion problem. Motion problems deal with a vehicle—usually a car or train—traveling some amount of time, at some speed, to cover some distance. There's a handy three-piece formula that you can use to solve these problems:

$$\textbf{Rate} \times \textbf{Time} = \textbf{Distance}$$

or

$$R \times T = D$$

If you have two of the three pieces, you can always plug into the formula to solve for the third.

I once saw a movie called *Planes, Trains, and Automobiles*. Maybe that will help me here.

Example: Lisa drives 5 hours and travels a distance of 225 miles. At what speed did Lisa drive?

Solution: Use the three-piece formula. You're told that Lisa drove 5 hours, so that is your time, or *T*. You're also told that she travels a distance of 225 miles, so that is your distance, or *D*. Since you have the time and the distance, you can solve for the rate of speed, *R*. Plug the values for *T* and *D* into the formula to solve for *R*:

$$R \times T = D$$
$$R \times 5 = 225$$
$$\frac{R \times 5}{5} = \frac{225}{5}$$
$$R = 45$$

So, Lisa drove at a speed of 45 miles per hour.

That was easy. By the way, who taught Lisa how to drive—her grandmother? Put the pedal to the metal, girl!

Get Wise!

Time to get in motion and solve the following motion problems.

1. The train trip from Penn Station to South Station takes 5.5 hours. If the train travels at a speed of 60 miles per hour, how many miles does it travel?

(A) $10.\overline{90}$ (D) 300

(B) 115 (E) 330

(C) 230

2. Doug drives 240 miles at 40 miles per hour. If he drove this distance at 60 miles an hour, how much less time would the trip take?

(A) 2 hours (D) 8 hours

(B) 4 hours (E) 10 hours

(C) 6 hours

3. A truck travels on the interstate highway for 5 hours at a speed of 65 miles per hour. It then travels on local roads for 4 hours, at a speed of 40 miles per hour. How many miles does the truck travel in total?

(A) 325 (D) 485

(B) 360 (E) 585

(C) 460

4. Meg drives x miles at y miles per hour. How long does this trip take, in terms of x and y?

(A) xy (D) $\dfrac{y}{x}$

(B) $x + y$ (E) $\dfrac{1}{xy}$

(C) $\dfrac{x}{y}$

All these trains and cars whizzing around—I'm getting all types of motion sickness!

5. Jane and Amar both make 1,050-mile trips by train. Jane takes the high-speed train, which travels at 150 miles an hour. Amar takes the regular train and his trip takes 8 hours longer than Jane's. How many miles per hour does the regular train travel?

(A) $58.\overline{33}$ (D) 131.25

(B) 70   (E) 150

(C) 75

How Wise?

Check your answers on page 163.

AVERAGE RATES

Some motion problems ask you to find the average rate (or average speed) for a trip driven at different rates.

So, it's like an average question and a motion question at the same time. I guess we need to multitask.

You don't handle this like a regular average problem, though. It's important to remember that you can't just average speeds. To find an average rate, you need to use the formula:

$$\text{Average Rate} = \frac{\text{Total Distance}}{\text{Total Time}}$$

Example: Jason lives 30 miles away from his friend Freddy. On Friday night, Jason drives to Freddy's house at a speed of 30 miles per hour. Jason loses track of the time and stays too long. He rushes to get home before his curfew, so that he doesn't make his mother angry. He drives home at a speed of 50 miles per hour. What is Jason's average speed, in miles per hour, for the round trip?

What a nightmare—not only math but curfews too?! What's next, questions about being grounded?

Solution: Again, use the formula: $\text{Average Rate} = \dfrac{\text{Total Distance}}{\text{Total Time}}$.

The common mistake a lot of people would make on this question would be to come up with 40 miles per hour as the average rate. This is the result of just averaging the two speeds:

$$\frac{30+50}{2} = \frac{80}{2} = 40$$

This is not the way to solve, though. You must use the average rate formula, and divide *total* distance by *total* time.

The total distance is easy. Jason drove 30 miles there and 30 miles back, for a total of 60 miles. But what about the total time? The question doesn't tell you how long either leg of the trip took. It does tell you the distance of each leg, though, and the rate at which Jason traveled it. Since you have two of the three pieces, you can use the formula $R \times T = D$ to solve for the third:

First Leg: $R \times T = D$

$$30 \times T = 30$$

$$\frac{30 \times T}{30} = \frac{30}{30}$$

$$T = \frac{1}{1} = 1$$

Second Leg: $R \times T = D$

$$50 \times T = 30$$

$$\frac{50 \times T}{50} = \frac{30}{50}$$

$$T = \frac{3}{5} = 0.6$$

So, the total time is $1 + 0.6$, or 1.6 hours. Now, you can plug into the average rate formula:

$$\text{Average Rate} = \frac{\text{Total Distance}}{\text{Total Time}}$$

$$\text{Average Rate} = \frac{60}{1.6}$$

$$\text{Average Rate} = 37.5$$

If you think about this answer, it makes sense. Since Jason spent more time driving at a speed of 30 miles per hour, his average speed should be closer to 30 than to 50.

WISE NOTE

The reasoning behind the average speed concept is similar to that behind the weighted average. Since one speed is driven for a longer amount of time, it should count more toward the total average.

Example: Irina drives from Town A to Town B at 60 miles per hour, and then returns to Town A at 40 miles per hour. What was her average speed for the entire trip, in miles per hour?

Town A and Town B? How creative! I'm surprised that Irina got a name and wasn't just called Girl 1. The people who write these math questions are so imaginative.

Solution: Make up a number for the distance.

This is another average rates question, so you just need to plug the total distance and total time into the formula. But this time you're not given any information about the distance between towns. Without this information, you can't figure out the amount of time for each leg of the trip either. So what do you do? Is this question unanswerable?

As much as I'm sure you'd like to have gotten out of it, there is a way to solve this question. Make up a number for the distance between the towns. Since this is a round trip, you know that the distance traveled on each leg of the trip is the same. Make up a number that will be easy to work with. Since one speed is 60 and one is 40, a good number would be one that is evenly divisible by both of them. Try 120. If each leg is 120 miles, the total distance is 240 miles. With a value for distance, you can figure out the amount of time for each leg:

$$\text{First Leg: } R \times T = D \qquad \text{Second Leg: } R \times T = D$$

$$60 \times T = 120 \qquad\qquad 40 \times T = 120$$

$$\frac{60 \times T}{60} = \frac{120}{60} \qquad\qquad \frac{40 \times T}{40} = \frac{120}{40}$$

$$T = 2 \qquad\qquad\qquad T = 3$$

So, the total time is 2 + 3 = 5 hours. Now, you can plug into the average rate formula:

$$\text{Average Rate} = \frac{\text{Total Distance}}{\text{Total Time}}$$

$$\text{Average Rate} = \frac{240}{5}$$

$$\text{Average Rate} = 48$$

If you think about this answer, it makes sense. It will always take more time to drive the same distance at a slower speed. This means that the average speed will be closer to the lower speed than the higher speed.

Here's something else that makes sense: It will always take more time to answer math questions when the people who write them make YOU come up with numbers instead of doing it themselves.

Get Wise!

If you can figure out the average rates problems below, you'll be better than average.

1. Tammy rides her bike 10 miles to work at a rate of 10 miles per hour. At the end of the day, she rides her bike home from work at a rate of 15 miles per hour. What was Tammy's average rate of speed for these two trips?

(A) 9.5   (D) 12.5

(B) 11.5   (E) 13

(C) 12

Poor Tammy—I know how it feels to not have a car. Sometimes life is so unfair!

2. Because of traffic, a truck driver made the 300-mile trip from Chicago to Milwaukee at a speed of 50 miles an hour. On the way back the roads were clear, and the truck driver was able to make the trip at 60 miles an hour. What was the truck driver's average rate of speed for the round-trip, in miles per hour?

(A) 52.5   (D) 57.25

(B) $54.\overline{54}$ (E) 58

(C) 55

3. Reese drove 90 miles from Centreville to Springfield, and then drove back to Centreville along the same route the next day. Her average rate of speed for this round-trip was 36 miles per hour. If she drove to Springfield at a speed of 45 miles per hour, what rate did she drive to Centreville, in miles per hour?

(A) 3

(B) 30

(C) 32.5

(D) 35

(E) 40.5

4. A plane makes a round-trip. On the way there, it travels at a speed of *a* miles per hour. On the way back, it travels at a speed of *b* miles per hour. If the length of each leg of the trip is *c*, what is the average speed for this round-trip in terms of *a*, *b*, and *c*?

(A) $\dfrac{a+b}{2}$

(B) $\dfrac{2c}{\dfrac{c}{a}+\dfrac{c}{b}}$

(C) $\dfrac{c}{a+b}$

(D) $\dfrac{\dfrac{2c}{a+b}}{c}$

(E) $2abc$

I've heard of planes making round-trips, but I've never heard of them moving at speeds of *a* and *b*.

5. Otto is running late, so he drives from his house to school at a speed
of 30 miles per hour. After school, he drives home at a speed of 20
miles per hour. What was Otto's average rate of speed for his round-
trip, in miles per hour?

(A) 22   (D) 25

(B) 23   (E) 26

(C) 24

How Wise?

Check your answers on page 165.

COMPLEX RATES PROBLEMS

Some complex questions talk about two different vehicles traveling at different rates of speed and ask when they meet. I'm sure you've heard questions like this. They're the kind of problems that send some people screaming in fear and dread. Don't freak out! Even though these problems are "complex," we'll give you a simple method to tackle them.

Example: An express train left the station 3.5 hours after a local train traveling in the same direction. The express train travels at a speed 30 miles less than three times the speed of the other train. If the express train catches up with the regular train in 2.5 hours, how many miles will the trains have traveled?

Here we go again, making something more confusing than it needs to be. Can't we just look at a train schedule or call someone at the train station to find out when the trains meet?

Solution: First, figure out what the question is asking. The express train catches up with the local train at the moment that the two trains meet. That means that the distance of the two trains will be the same at this point. This fact will help you solve the problem. There's a lot going on in this problem, so it will help if you organize it in a chart. For example:

Train	Rate	Time	Distance
Express			
Local			

This chart has a row for each train, and a column for each of the three pieces of the motion formula. You're not given the rate of the express train, but you're told that it's 30 miles less than three times the speed of the regular train.

If you let the speed of the local train be R, the express train's rate is $3R - 30$. Since the local train left 3.5 hours before the express train, it has traveled for a total of $2.5 + 3.5 = 6$ hours when the trains meet.

Train	Rate	Time	Distance
Express	3R – 30	2.5	2.5(3R – 30)
Local	R	2.5 + 3.5 = 6	6R

You want to find the distance when the two trains meet, at which point both will have traveled the same number of miles. Remember, Rate × Time = Distance, so you can multiply the second and third columns to find an expression for the distance in each row. The trains will have traveled the same number of miles when they meet, so you can set these two expressions equal and solve for R:

$$2.5(3R - 30) = 6R$$
$$7.5R - 75 = 6R$$
$$-75 = -1.5R$$
$$\frac{-75}{-1.5} = \frac{-1.5R}{-1.5}$$
$$50 = R$$

Now that you have the value of R, you can plug it into the distance expression for either train to solve. The expression for the local train is much simpler, so use that:

$$6R = \text{Distance}$$
$$6(50) = \text{Distance}$$
$$300 = \text{Distance}$$

Get Wise!

Now, see if you can ace these complex rates problems.

1. Kaneesha leaves Los Angeles at 1 p.m. and drives north on Interstate 5 at a speed of 50 miles per hour. Justin leaves Los Angeles at 3 p.m. and drives north on Interstate 5 at a speed of 65 miles per hour. At what time will Justin catch up with Kaneesha?

(A) 5:40 p.m.   (D) 9:40 p.m.

(B) 6:40 p.m.   (E) 12:40 a.m.

(C) 7:40 p.m.

I don't know Kaneesha, and I don't know Justin. Why do I need to know when and where they meet?

2. Two trains leave at the same time from stations that are 560 miles apart, traveling toward each other. The rate of Train 1 is 10 miles less than twice the rate of Train 2. If the trains pass each other after 4 hours, how fast was Train 1 traveling, in miles per hour?

(A) 50   (D) 90

(B) 70   (E) 100

(C) 80

3. A freight train leaves the station 3 hours before a passenger train traveling in the same direction. The rate of the freight train is 30 miles an hour less than that of the passenger train. If the passenger train catches up with the freight train after 2 hours, how many miles will the trains have traveled?

(A) 20   (D) 90

(B) 50   (E) 100

(C) 70

4. A car and a truck leave at the same time from destinations 510 kilo-meters apart, traveling toward each other. The rate of the truck is 30 kilometers per hour more than that of the car. If the truck and car pass each other after 3 hours, how fast was the truck traveling, in kilometers per hour?

(A) 30   (D) 100

(B) 40   (E) 210

(C) 70

I'm glad that they passed each other. For a minute there I was worried that we were going to have a collision on our hands.

5. Jamie left home at 8 a.m., traveling along the highway at 40 miles per hour. At 11 a.m., Josette left home and started after her, traveling on the same road at 60 miles per hour. At what time did Josette catch up with Jamie?

(A) 2 p.m.   (D) 8 p.m.

(B) 5 p.m.   (E) 11 p.m.

(C) 6 p.m.

How Wise?

Check your answers on page 168.

A Word to the Wise

* For motion problems, use the three-piece formula: Rate × Time = Distance. If you have two of the three pieces, you can always solve for the third.

* You can't just average two speeds. To find the average rate, use the formula: $\text{Average Rate} = \dfrac{\text{Total Distance}}{\text{Total Time}}$

* Some complex questions talk about two different vehicles traveling at different rates of speed, and ask when they meet. To answer these questions, organize the information you are given in a chart. Use a variable to represent an unknown value and write an equation to solve for its value.

ANSWERS TO CHAPTER 7: PRACTICE EXERCISES

Motion Problems (Page 149)

1. **The correct answer is (E).** Use the three-piece formula. Time is 5.5 and the rate is 60. Plug into the formula and solve for the distance:

$$R \times T = D$$
$$60 \times 5.5 = D$$
$$330 = D$$

2. **The correct answer is (A).** Use the three-piece formula. Doug drives 240 miles at 40 miles per hour, so $D = 240$ and $R = 40$. Plug into the formula and solve for the time:

$$R \times T = D$$
$$40 \times T = 240$$
$$T = 6$$

Now, find the time if Doug had driven 240 miles at 60 miles per hour:

$$60 \times T = 240$$
$$T = 4$$

So, it would have taken $6 - 4 = 2$ hours less if Doug had driven at 60 miles per hour.

3. **The correct answer is (D).** Use the three-piece formula. The truck travels for 5 hours at 65 miles per hour, so $T = 5$ and $R = 65$. Plug into the formula and solve for the distance:

$$R \times T = D$$
$$65 \times 5 = D$$
$$325 = D$$

Then the truck travels for 4 hours at 40 miles per hour, so $T = 4$ and $R = 40$. Plug in and solve for the distance:

$$40 \times 4 = D$$
$$160 = D$$

The truck travels a total of $325 + 160 = 485$ miles.

4. **The correct answer is (C).** Use the three-piece formula. Meg drives a total of x miles at a rate of y miles per hour, so $D = x$ and $R = y$. Plug into the formula and solve for the time:

$$R \times T = D$$
$$y \times T = x$$
$$\frac{y \times T}{y} = \frac{x}{y}$$
$$T = \frac{x}{y}$$

5. **The correct answer is (B).** Use the three-piece formula. Jane travels a total of 1,050 miles at a rate of 150 miles per hour, so $D = 1,050$ and $R = 150$. Plug into the formula and solve for the time:

$$R \times T = D$$
$$150 \times T = 1,050$$
$$\frac{150 \times T}{150} = \frac{1,050}{150}$$
$$T = 7$$

Amar also travels a total of 1,050 miles, but his trip takes 8 hours longer, or $7 + 8 = 15$ hours. Plug $D = 1,050$ and $T = 15$ into the formula to solve for R:

$$R \times 15 = 1,050$$
$$\frac{R \times 15}{15} = \frac{1,050}{15}$$
$$R = 70$$

Average Rates Problems (Page 155)

1. **The correct answer is (C).** Use the average rate formula. Tammy travels the 10-mile trip twice, so her total distance is 2(10) = 20. To figure out the total time, use the motion formula to figure out the time for each trip:

$$R \times T = D \qquad\qquad R \times T = D$$
$$10 \times T = 10 \qquad\qquad 15 \times T = 10$$
$$T = \frac{10}{10} = 1 \qquad\qquad T = \frac{10}{15} = \frac{2}{3}$$

Now you are ready to plug into the average rate formula:

$$\text{Average Rate} = \frac{\text{Total Distance}}{\text{Total Time}}$$

$$\text{Average Rate} = \frac{20}{1 + \frac{2}{3}}$$

$$\text{Average Rate} = \frac{20}{\frac{5}{3}}$$

$$\text{Average Rate} = 20 \times \frac{3}{5} = 12$$

2. **The correct answer is (B).** Use the average rate formula. The trucker makes the 300-mile trip twice, for a total distance of 2(300) = 600 miles. To figure out the total time, use the motion formula to figure out the time for each trip:

$$R \times T = D \qquad\qquad R \times T = D$$
$$50 \times T = 300 \qquad\qquad 60 \times T = 300$$
$$T = \frac{300}{50} = 6 \qquad\qquad T = \frac{300}{60} = 5$$

Now you are ready to plug into the average rate formula:

$$\text{Average Rate} = \frac{\text{Total Distance}}{\text{Total Time}}$$

$$\text{Average Rate} = \frac{600}{6+5}$$

$$\text{Average Rate} = \frac{600}{11}$$

$$\text{Average Rate} = 54.\overline{54}$$

3. **The correct answer is (B).** Use the average rate formula. Reese makes the 90-mile trip twice, so her total distance is 2(90) = 180. You're given that the first leg of her trip was driven at 45 miles per hour, so you can use the motion formula to figure out the time it took:

$$R \times T = D$$
$$45 \times T = 90$$
$$T = 2$$

You have the total distance, the average rate, and part of the time. Let the rest of the time be t, and plug all this information into the average rate formula to solve for t:

$$\text{Average Rate} = \frac{\text{Total Distance}}{\text{Total Time}}$$

$$36 = \frac{180}{2+t}$$

$$36(2+t) = 180$$

$$72 + 36t = 180$$

$$36t = 108$$

$$t = 3$$

Since you have the time for the second leg and the distance, you can plug into the motion formula to solve for the rate:

$$R \times T = D$$

$$R \times 3 = 90$$

$$R = 30$$

4. **The correct answer is (B).** Use the average rate formula. The plane makes the c-mile trip twice, for a total distance of $2c$. To figure out the total time, use the motion formula to figure out the time for each trip:

$$R \times T = D \qquad\qquad R \times T = D$$

$$a \times T = c \qquad\qquad b \times T = c$$

$$\frac{a \times T}{a} = \frac{c}{a} \qquad\qquad \frac{b \times T}{b} = \frac{c}{b}$$

$$T = \frac{c}{a} \qquad\qquad\quad T = \frac{c}{b}$$

Now you are ready to plug into the average rate formula:

$$\text{Average Rate} = \frac{\text{Total Distance}}{\text{Total Time}}$$

$$\text{Average Rate} = \frac{2c}{\dfrac{c}{a} + \dfrac{c}{b}}$$

5. **The correct answer is (C).** Use the average rate formula. You're not given a value for the distance of each leg of the round-trip, so make one up. Use a number that's evenly divisible by both rates, say 60. This makes the total distance $2(60) = 120$. To figure out the total time, use the motion formula to figure out the time for each trip:

$$R \times T = D \qquad\qquad R \times T = D$$
$$30 \times T = 60 \qquad\quad 20 \times T = 60$$
$$T = 2 \qquad\qquad\quad T = 3$$

Now you are ready to plug into the average rate formula:

$$\text{Average Rate} = \frac{\text{Total Distance}}{\text{Total Time}}$$

$$\text{Average Rate} = \frac{120}{2+3}$$

$$\text{Average Rate} = \frac{120}{5}$$

$$\text{Average Rate} = 24$$

Complex Rates Problems (Page 160)

1. **The correct answer is (D).** Put the information in a chart:

	Rate	Time	Distance
Kaneesha	50	$T + 2$	$50(T + 2)$
Justin	65	T	$65T$

You want to find the distance when the two meet, at which point both will have traveled the same number of miles. Since Rate × Time = Distance, you can multiply through the first two columns to find an expression for the distance in each row. Set these two expressions equal and solve for T:

$$50(T + 2) = 65T$$
$$50T + 100 = 65T$$
$$100 = 15T$$
$$\frac{100}{15} = T$$
$$6\frac{2}{3} = T$$

This means that Justin will have traveled for $6\frac{2}{3}$ hours, or 6 hours and 40 minutes when he catches up. Since he left home at 3 p.m., he catches up with Kaneesha at 9:40 p.m.

2. **The correct answer is (D).** Put the information in a chart.

	Rate	Time	Distance
Train 1	$2R - 10$	4	$4(2R - 10)$
Train 2	R	4	$4R$

You want to find the distance when the two trains pass each other, at which point they will have traveled a total of 560 miles (the distance between them). Since Rate × Time = Distance, you can multiply through the first two columns to find an expression for the distance in each row. Write an equation summing these expressions to 560 and solve for R:

$$4(2R - 10) + 4R = 560$$
$$8R - 40 + 4R = 560$$
$$12R - 40 = 560$$
$$12R = 600$$
$$R = 50$$

The rate of Train 1 is $2R - 10$. Plugging in $R = 50$, you find that its rate is: $2(50) - 10 = 100 - 10 = 90$.

3. **The correct answer is (E).** Put the information in a chart.

	Rate	Time	Distance
Freight train	$R - 30$	$2 + 3 = 5$	$5(R - 30)$
Passenger train	R	2	$2R$

You want to find the distance when the two trains meet, at which point both will have traveled the same number of miles. Since Rate × Time = Distance, you can multiply through the first two columns to find an expression for the distance in each row. Set these two expressions equal and solve for R:

$$5(R - 30) = 2R$$
$$5R - 150 = 2R$$
$$-150 = -3R$$
$$50 = R$$

Plug this value for R into either distance expression to solve. Use that of the passenger train since it's simpler: $2R = 2(50) = 100$.

4. **The correct answer is (D).** Put the information in a chart.

	Rate	Time	Distance
Car	R	3	$3R$
Truck	$R + 30$	3	$3(R + 30)$

You want to find the distance when the car and truck pass each other, at which point they will have traveled a total of 510 kilometers (the distance between them). Since Rate × Time = Distance, you can multiply through the first two columns to find an expression for the distance in each row. Write an equation summing these expressions to 510 and solve for R:

$$3R + 3(R + 30) = 510$$
$$3R + 3R + 90 = 510$$
$$6R + 90 = 510$$
$$6R = 420$$
$$R = 70$$

The rate of the truck is $R + 30$, or $70 + 30 = 100$.

5. **The correct answer is (B).** Put the information in a chart.

	Rate	Time	Distance
Jamie	40	T + 3	40(T + 3)
Josette	60	T	60T

You want to find the distance when the two meet, at which point both will have traveled the same number of miles. Since Rate × Time = Distance, you can multiply through the first two columns to find an expression for the distance in each row. Set these two expressions equal and solve for T:

$$40(T + 3) = 60T$$
$$40T + 120 = 60T$$
$$120 = 20T$$
$$6 = T$$

This means that Josette traveled for 6 hours. Since she left home at 11 a.m., she catches up with Jamie at 5 p.m.

chapter 8

Work It

UNDERSTANDING WORK PROBLEMS

Work problems? Aren't all the problems in this book work? I've got a bad feeling about this chapter . . .

Actually, Chi, we think you'll like this chapter. It may come in handy the next time your parents ask you to do a job around the house. Use your knowledge of work problems and you just might be able to persuade them that it's best if you don't do it alone.

If working through this chapter can save me some work around the house, I'm willing to read on.

Example: It takes Chi 5 hours to clean her room by herself. It only takes her mother 2 hours to clean the entire room by herself. How long will it take if Chi and her mother clean her room together?

Solution: Take a minute to think about how big the answer should be. Clearly working together will take less time than either person working alone. Since Chi's mom can clean the room in 2 hours, it should take the two of them together less than 2 hours. Thinking through to a ballpark idea of the answer could help you eliminate incorrect answers on a multiple-choice question. In this case, you could eliminate any answers that were not less than 2.

First, figure out how much work each person can do per hour. It takes Chi 5 hours to clean the room alone, so she does $\frac{1}{5}$ of the job each hour. Since her mom can do the whole job in 2 hours, she does $\frac{1}{2}$ of the job each hour. Add these rates to see how much of the job they do together each hour:

$$\frac{1}{5} + \frac{1}{2} = \frac{2}{10} + \frac{5}{10} = \frac{7}{10}$$

Working together, Chi and her mom can do $\frac{7}{10}$ of the job each hour. For the job to be done, the total amount of work must equal 1. If you let x be the number of hours it takes to complete the job, you can multiply this by the rate $\frac{7}{10}$, set this equal to 1, and figure out how long it takes for both Chi and her mom to do the cleaning.

$$\frac{7}{10}x = 1$$

$$\frac{\frac{7}{10}x}{\frac{7}{10}} = \frac{1}{\frac{7}{10}}$$

$$x = \frac{10}{7}$$

So, it takes $\frac{10}{7}$ or $1\frac{3}{7}$ hours for them to clean the room together. Actually, as soon as you found out that they do $\frac{7}{10}$ of the job each hour, you know that the total job takes $\frac{10}{7}$ hours. Since the reciprocal of the total time to do the job equals the hourly rate, the reciprocal of the hourly rate equals the total amount of time to do the job. In other words, since you were able to say that Chi does $\frac{1}{5}$ of the job each hour, because it takes her 5 hours to do the whole job, you can say that since she and her mom do $\frac{7}{10}$ of the job each hour, it takes them $\frac{10}{7}$ hours to do the whole job.

Working together sounds better for everyone. I like the fact that I'll spend so much less time cleaning. And, my mom's always saying that we should do more things together.

Example: It takes Tatiana *t* hours to complete a job by herself, and it takes Jacob *j* hours to complete the same job by himself. If they work together, how long will it take them to complete the job, in terms of *t* and *j*?

Solution: Don't get freaked out by all the variables. It's the same type of problem as we just looked at, aside from the fact that variables are substituted for numbers. Handle it the same way.

First, figure out how much work each person can do per hour. Tatiana takes *t* hours to do the job by herself, so she does $\frac{1}{t}$ of the job each hour. Jacob takes *j* hours to do the job by himself, so he does $\frac{1}{j}$ of the job each hour.

Add these rates to see how much of the job they do together each hour. Remember, to find a common denominator for two fractions, you can multiply the numerator and denominator of each fraction by the denominator of the other. Therefore:

$$\frac{1}{t} + \frac{1}{j} = \frac{(j)1}{(j)t} + \frac{(t)1}{(t)j} = \frac{j}{jt} + \frac{t}{jt} = \frac{j+t}{jt}$$

Remember, the reciprocal of the hourly rate is the total number of hours necessary to complete the job. Since they do $\frac{j+t}{jt}$ of the job each hour, together it takes Tatiana and Jacob $\frac{jt}{j+t}$ hours to complete the job.

WISE NOTE

Had this last problem been a multiple-choice question, there's another way you could have solved it. If the sight of the variables froze you in your tracks, you could have made up numbers for them. Then, you would solve the problem with the numbers and come up with a numerical answer. Finally, you'd plug your "made up" numbers for the variables into the answer choices to see which matched the answer you came up with. To read about this strategy in more detail, check out Chapter Ten, "So Many Variables, So Little Time."

Get Wise!

Time to "work it" on your own. Try the following work problems.

1. Working alone, Will can finish a job in 6 hours. Working alone, Grace can finish the same job in 10 hours. If they work together, how long will it take them to do this job?

(A) $\dfrac{4}{15}$

(D) 4

(B) $1\dfrac{1}{3}$

(E) 8

(C) $3\dfrac{3}{4}$

2. Two pipes are being used to drain the water from the community swimming pool. The first pipe drains water at a rate of x gallons per hour. The second pipe drains water at a rate of y gallons per hour. With the two pipes working together, how long will it take for the pool to be emptied?

(A) $x + y$

(D) $\dfrac{xy}{x + y}$

(B) $\dfrac{x + y}{2}$

(E) $\dfrac{1}{x + y}$

(C) $\dfrac{x + y}{xy}$

3. It takes Frank 5 hours to paint a room by himself. It takes Hildy 4 hours to paint the same room by herself. If they work together, how long will it take them to paint the room together?

(A) $\dfrac{9}{20}$ (D) $3\dfrac{4}{5}$

(B) $2\dfrac{2}{9}$ (E) $4\dfrac{1}{2}$

(C) $2\dfrac{1}{2}$

I'd love to give my room a makeover. How hard could painting be? The problem would be getting my parents to pay for all the cool stuff I'd need.

4. It takes Henry n hours to rake the leaves in the yard alone. It takes Cole twice as long to rake the leaves alone. How many hours will it take Henry and Cole to rake the yard if they work together?

(A) $2n$ (D) $\dfrac{n+2}{2}$

(B) $\dfrac{n}{2}$ (E) $\dfrac{2n}{3}$

(C) $\dfrac{n}{3}$

5. A standard printer can complete a certain print job in 9 hours. A high-speed printer can complete the same job in 5 hours. How many hours would it take to complete the print job if two standard printers and one high-speed printer were run simultaneously?

(A) $1\dfrac{22}{23}$　　　　(D) 4

(B) $2\dfrac{7}{19}$　　　　(E) 7

(C) $3\dfrac{3}{14}$

How Wise?

Check your answers on page 186.

SOLVING FOR AN INDIVIDUAL RATE

In the problems we've looked at so far, you were given two individual rates and asked to solve for the combined rate. Some questions will give you one individual rate and the combined rate, and ask you to solve for the other individual rate. Don't be intimidated by these questions—you use the same basic method to solve.

Example: Working together, two machines in a factory can fill an urgent order for widgets in 5 hours. Working alone, the first machine can fill the entire order in 9 hours. How long would it take the second machine, working alone, to fill the entire order?

Widgets? What are widgets and why would someone have an urgent need for them?

Solution: Take a minute to think about how big the answer should be. Working together the machines will take less time to complete the job than either machine would take working alone. Since the two machines together take 5 hours to complete the job, each machine working alone should take more than 5 hours to do the job. If this were a multiple-choice question, you could eliminate any answers that were less than 5.

Now, work with the information you're given to figure out the hourly rates. Since the first machine working alone takes 9 hours to complete the job, it does $\frac{1}{9}$ of the job each hour. We don't know how many hours it takes the second machine to do the job alone. That's what we're trying to figure out. Use a variable to represent this number, say x. If it takes the second machine x hours to do the job alone, it does $\frac{1}{x}$ of the job each hour. Makes sense? You also have information about the combined rate. Working together it takes the two machines 5 hours to complete the job. That means that each hour $\frac{1}{5}$ of the job is getting done.

As we did in the previous problem, we add the individual rates to get the combined rate. The difference is that this time we have one individual rate and the combined rate. Therefore:

$$\frac{1}{9} + \frac{1}{x} = \frac{1}{5}$$

Isolate the fraction with the variable on one side of the equation. Therefore:

$$\frac{1}{9} + \frac{1}{x} - \frac{1}{9} = \frac{1}{5} - \frac{1}{9}$$

$$\frac{1}{x} = \frac{9}{45} - \frac{5}{45}$$

$$\frac{1}{x} = \frac{4}{45}$$

Now you can cross-multiply to solve for x. Or, you could just take the reciprocal of each side of the equation to solve for x. Either way, the value of x is $x = \frac{45}{4}$, or $11\frac{1}{4}$.

You'd think that by now we would have solved for x for good.

Get Wise!

Flex those muscles again—it's time for more work problems. This time, you have to solve for an individual rate.

1. Working together, Vern and Lori can tile a kitchen floor in 6 hours. If it takes Lori 10 hours to tile the kitchen by herself, how many hours does it take Vern to tile it by himself?

(A) $3\dfrac{3}{4}$ (D) 15

(B) 4   (E) 16

(C) 8

2. You bring your car to the garage to have the transmission fixed. Two mechanics working together will complete the job in 7 hours. If it takes one of the mechanics 12 hours to do the job alone, how many hours does it take the other mechanic to do the job alone?

(A) $4\dfrac{8}{19}$ (D) 19

(B) 5   (E) $24\dfrac{7}{12}$

(C) $16\dfrac{4}{5}$

3. Working together, Computers 1 and 2 take $6\frac{2}{3}$ hours to back up the data for a week's worth of work. If Computer 1 takes 15 hours to complete the job alone, how many hours does it take Computer 2 to do the job alone?

(A) $4\frac{8}{13}$ (D) $21\frac{2}{3}$

(B) $8\frac{1}{3}$ (E) 30

(C) 12

4. Greg and Larry are working together to sand and refinish the hardwood floor in a room. It takes them 18 hours to complete the job working together. If it takes Greg 30 hours to do the job alone, how many hours does it take Larry to do the job alone?

(A) 12   (D) 54

(B) 45   (E) 60

(C) 48

5. Julie and Maureen are quilting a bedspread. Working together, they will finish quilting in 9 hours. If it takes Maureen 21 hours to do the quilting by herself, how many hours does it take Julie to do the quilting by herself?

(A) 12   (D) 30

(B) $15\frac{3}{4}$ (E) $21\frac{3}{7}$

(C) 23

Quilting? Who is this book for—me or my grandmother?

How Wise?

Check your answers on page 187.

A Word to the Wise

* The reciprocal of the total time it takes to complete a job is equal to the hourly rate.

* The reciprocal of the hourly rate is equal to the total time it takes to complete the job.

* The sum of the individual rates is the combined rate. Use this fact to set up your problems and solve for the unknown value.

* If a multiple-choice work problem uses variables instead of numerical values, you can make up numbers for those variables. Solve the problem with these numbers. Then, plug these numbers into the choices to see which matches.

ANSWERS TO CHAPTER 8: PRACTICE EXERCISES

Work Problems (Page 178)

1. **The correct answer is (C).** Working together, Will and Grace can finish the job in less time than it would take either one alone. Since Will can do the job alone in 6 hours, eliminate answer choice (E). Add the individual hourly rates to find the combined hourly rate: $\frac{1}{6} + \frac{1}{10} = \frac{5}{30} + \frac{3}{30} = \frac{8}{30}$. The reciprocal of this sum is the total number of hours it takes for the two working together. Therefore: $\frac{30}{8} = 3\frac{3}{4}$.

2. **The correct answer is (D).** Add the individual hourly rates to find the combined hourly rate: $\frac{1}{x} + \frac{1}{y} = \frac{y}{xy} + \frac{x}{xy} = \frac{x+y}{xy}$. The reciprocal of this sum is the total number of hours it takes for the two pipes working together. Therefore: $\frac{xy}{x+y}$.

3. **The correct answer is (B).** Add the individual hourly rates to find the combined hourly rate: $\frac{1}{5} + \frac{1}{4} = \frac{4}{20} + \frac{5}{20} = \frac{9}{20}$. The reciprocal of this sum is the total number of hours it takes for Frank and Hildy working together. Therefore: $\frac{20}{9} = 2\frac{2}{9}$.

4. **The correct answer is (E).** Since it takes Cole twice as long as Henry to rake, it takes Cole $2n$ hours to rake alone. Add the individual hourly rates to find the combined hourly rate: $\frac{1}{n} + \frac{1}{2n} = \frac{2}{2n} + \frac{1}{2n} = \frac{3}{2n}$. The reciprocal of this sum is the total number of hours it takes for the two working together. Therefore: $\dfrac{2n}{3}$.

5. **The correct answer is (B).** Add the individual hourly rates to find the combined hourly rate. Remember, there are two standard printers and one high-speed printer being used:

$$\frac{1}{9}+\frac{1}{9}+\frac{1}{5}=\frac{5}{45}+\frac{5}{45}+\frac{9}{45}=\frac{19}{45}.$$

The reciprocal of this sum is the total number of hours it takes for the two working together. Therefore: $\dfrac{45}{19}=2\dfrac{7}{19}$.

Solving for an Individual Rate (Page 183)

1. **The correct answer is (D).** Let x be the number of hours it takes Vern to tile the floor alone. Add the individual hourly rates and isolate the fraction with the variable on one side of the equation:

$$\frac{1}{10}+\frac{1}{x}=\frac{1}{6}$$
$$\frac{1}{x}=\frac{1}{6}-\frac{1}{10}$$
$$\frac{1}{x}=\frac{5}{30}-\frac{3}{30}$$
$$\frac{1}{x}=\frac{2}{30}$$

Solve for x by taking the reciprocal of both sides of the equation:

$$x=\frac{30}{2}=15$$

2. **The correct answer is (C).** Working together, the two mechanics complete the job in less time than either would alone. Since they take 7 hours to complete the job together, each working alone should take more than 7 hours. Eliminate answer choices (A) and (B) for being too small. Let x be the number of hours it takes the other mechanic to do the job alone. Add the individual hourly rates and isolate the fraction with the variable on one side of the equation:

$$\frac{1}{12}+\frac{1}{x}=\frac{1}{7}$$
$$\frac{1}{x}=\frac{1}{7}-\frac{1}{12}$$
$$\frac{1}{x}=\frac{12}{84}-\frac{7}{84}$$
$$\frac{1}{x}=\frac{5}{84}$$

Solve for x by taking the reciprocal of both sides of the equation:

$$x=\frac{84}{5}=16\frac{4}{5}$$

3. **The correct answer is (C).** Working together, Computers 1 and 2 complete the job in less time than either would alone. Since they take $6\frac{2}{3}$ hours to complete the job together, each working alone should take more than $6\frac{2}{3}$ hours. Eliminate answer choice (A) for being too small. Let x be the number of hours it takes Computer 2 to do the job alone. Add the individual hourly rates and isolate the fraction with the variable on one side of the equation:

$$\frac{1}{15} + \frac{1}{x} = \frac{1}{6\frac{2}{3}}$$

$$\frac{1}{x} = \frac{1}{\frac{20}{3}} - \frac{1}{15}$$

$$\frac{1}{x} = \frac{3}{20} - \frac{1}{15}$$

$$\frac{1}{x} = \frac{9}{60} - \frac{4}{60}$$

$$\frac{1}{x} = \frac{5}{60}$$

Solve for x by taking the reciprocal of both sides of the equation:

$$x = \frac{60}{5} = 12$$

4. **The correct answer is (B).** Working together, Greg and Larry complete the job in less time than either would alone. Since they take 18 hours to complete the job together, each working alone should take more than 18 hours. Eliminate answer choice (A) for being too small. Let x be the number of hours it takes Larry to complete the job alone. Add the individual hourly rates and isolate the fraction with the variable on one side of the equation:

$$\frac{1}{30} + \frac{1}{x} = \frac{1}{18}$$

$$\frac{1}{x} = \frac{1}{18} - \frac{1}{30}$$

$$\frac{1}{x} = \frac{5}{90} - \frac{3}{90}$$

$$\frac{1}{x} = \frac{2}{90}$$

Solve for x by taking the reciprocal of both sides of the equation:

$$x = \frac{90}{2} = 45$$

5. **The correct answer is (B).** Let x be the number of hours it takes Julie to do the job alone. Add the individual hourly rates and isolate the fraction with the variable on one side of the equation:

$$\frac{1}{21} + \frac{1}{x} = \frac{1}{9}$$

$$\frac{1}{x} = \frac{1}{9} - \frac{1}{21}$$

$$\frac{1}{x} = \frac{7}{63} - \frac{3}{63}$$

$$\frac{1}{x} = \frac{4}{63}$$

Solve for x by taking the reciprocal of both sides of the equation:

$$x = \frac{63}{4} = 15\frac{3}{4}$$

chapter 9

Mix It Up

UNDERSTANDING MIXTURE PROBLEMS

Mixture problems often involve scientists blending different solutions. Sometimes you're given specific solutions (saline, acid, and so on) and sometimes it's more mysterious (Solution A, Solution B, and so on). You'll be asked to figure out how much of a particular solution is required, or what percent strength the resulting blend will be.

Hey, maybe you should check the title of this book. It's math, not science. How much am I supposed to deal with at the same time?

Don't let the science scare you off. Just disregard the situation of the question and focus on the math. These questions can just as easily talk about mixing up a batch of fruit punch or paint color.

Example: A mad scientist needs a 60% solution for an experiment. How many liters of a 75% acid solution does he have to add to 30 liters of a 40% acid solution to produce a 60% acid solution?

Solution: Set up a chart to organize the information you are given. In many word problems, you might have a general sense of what to do. Don't feel bad if you don't have any idea of how to get started on this one. We'll give you a very structured method to set up and solve these questions. Once you've seen how to do it, these questions won't be so scary.

You're given a lot of information in this question, so organizing it in a chart would be helpful. It should look something like this:

Solution	% of Acid	Liters	Total Liters Acid
Solution 1			
Solution 2			
Mixture			

This chart gives you a place to record the different information about each of the different solutions. Now you can start filling in. The percent of acid in each solution is given, so you can write these down in the first blank column. Write them as decimals, since you'll be doing some multiplication later. You're asked to find the number of liters of the 75% solution, so let x represent that number. You're told that there are 30 liters of the 40% percent solution, so fill that slot in as well. How many liters are there of the mixture? Since this mixture is made by combining the other two solutions, it is the sum of these solutions. That means that you can fill $x + 30$ in its slot, as shown here.

Solution	% of Acid	Liters	Total Liters Acid
Solution 1	.75	x	
Solution 2	.40	30	
Mixture	.60	$x + 30$	

To find the total liters acid for each mixture, multiply together the percent of acid and the number of liters in each row, as shown here.

Solution	% of Acid	Liters	Total Liters Acid
Solution 1	.75	x	.75x
Solution 2	.40	30	.40(30)
Mixture	.60	$x + 30$	.60($x + 30$)

Next, use the values in this last column to set up an equation. Since the two solutions are combined together to form the mixture, the expressions for total liters acid for the two solutions is equal to the total liters acid of the mixture. Solve the equation for x:

$$.75x + .40(30) = .60(x + 30)$$
$$.75x + 12 = .6x + 18$$
$$.75x = .6x + 6$$
$$.15x = 6$$
$$\frac{.15x}{.15} = \frac{6}{.15}$$
$$x = 40$$

The mad scientist needs 40 liters of the 75% solution. If you think about this, it makes sense. Since 75% is closer to 60% than 40%, it makes sense that you'd need more liters of the 75% solution.

The whole thing sounds ridiculous to me. If the mad scientist had just bought 60% solution in the first place, it would have been a lot less work for all of us.

In this first example, you were given the number of liters of one of the solutions. What if you aren't told how many liters of either solution you have? Let's look at an example and find out.

Example: Ellen is catering a party and plans to mix together two fruit beverages to make punch. One of the beverages is 70% fruit juice and the other is 40% fruit juice. How many liters of each beverage should be used to create 75 liters of punch that is 50% fruit juice?

Solution: Don't worry that this question asks you to find two amounts instead of one. You still set up and solve using the same method we just covered.

 **One problem that makes you come up with two answers? That's not fair. It's kind of like this whole chapter, making me do science and math at the same time.**

First, set up your chart. While some of the labels have changed, notice that this is the same type of chart used in the previous solution. Fill in the pieces of information that are directly given in the question.

Beverage	% of Juice	Liters	Total Liters Juice
Beverage 1	.70		
Beverage 2	.40		
Punch	.50	75	

How do you fill in the liters for the two beverages? Well, let x be the number of liters of Beverage 1. That's simple enough. But how do you represent the number of liters of Beverage 2? Since you know that there are a total of 75 liters of the two beverages, you can subtract the amount of one beverage from the whole to find the number of liters of the other. Since you used x to represent

Beverage 1, there are $75 - x$ liters of Beverage 2. Now that you have the percent and liters columns filled in, you can multiply them to find the total liters for each row.

Beverage	% of Juice	Liters	Total Liters Juice
Beverage 1	.70	x	$.70x$
Beverage 2	.40	$75 - x$	$.40(75 - x)$
Punch	.50	75	$.50(75)$

Now you can set up an equation to solve for x:

$$.70x + .40(75 - x) = .50(75)$$
$$.70x + 30 - .40x = 37.5$$
$$.30x + 30 = 37.5$$
$$.30x = 7.5$$
$$\frac{.30x}{.30} = \frac{7.5}{.30}$$
$$x = 25$$

This means that you need 25 liters of the 70% beverage and $75 - 25 = 50$ liters of the 40% beverage. It makes sense that you would need more liters of the 40% beverage, since 40% is closer to 50% than 70% is.

WISE NOTE

Most mixture problems give you the percents of the different solutions and ask you to solve for one or more amounts. Some questions could give you the amounts and ask you to solve for a percent. You follow the same method: fill the information you're given into the chart, label the unknown x, and set up an equation to solve for its value.

Get Wise!

Now you can mix it up with the following mixture problems.

1. How many liters of a 30% saline solution must be added to 12.5 liters of a 10% saline solution to produce a solution that is 25% saline?

   (A) 7.5   **(D)** 37.5

   (B) 17.5   **(E)** 50

   (C) 25

2. Doug needs 40 gallons of paint that is 25% pigment. He has some leftover paint that is 10% pigment, and some that is 50% pigment. How many gallons of each must he use to mix the paint that he needs?

   (A) 20 gallons of 10%, 20 gallons of 50%

   (B) 25 gallons of 10%, 15 gallons of 50%

   (C) 15 gallons of 10%, 25 gallons of 50%

   (D) 30 gallons of 10%, 10 gallons of 50%

   (E) 10 gallons of 10%, 30 gallons of 50%

Hey, Doug, I'll make you an offer. If you go to the store and just buy the paint you need, I'll pay for it. It may cost me a month of allowances, but I'll make the sacrifice to avoid one of these sneaky questions that make you find two answers.

3. How many liters of a 20% fruit juice must be mixed with 15 liters of a 50% fruit juice to create a fruit punch that is 35% juice?

(A) 15   (D) 40
(B) 20   (E) 45
(C) 30

4. A 25% acid solution is created by combining 18 liters of Solution A with 6 liters of Solution B. If the acidity of Solution A is 30%, what is the acidity of Solution B?

(A) 5%   (D) 27.5%
(B) 10%   (E) 55%
(C) 15%

5. How many liters of an 80% saline solution must be added to 30 gallons of a 10% saline solution to create a solution that is 50% saline?

(A) 10   (D) 60
(B) 20   (E) 80
(C) 40

How Wise?

Check your answers on page 205.

OTHER MIXTURE PROBLEMS

In the problems we've looked at so far, whether they dealt with acidic solutions or fruit punches, the solutions were identified by the percent strength they had. Now we'll look at some mixture problems dealing with items that are identified by their prices. Since it's unlikely that you spend your spare time mixing chemical solutions or gallons of paint, the following questions may seem more realistic and applicable to your daily life.

Well, we'll see about that. Unless these questions involve mixing and matching outfits, they'll still be pretty irrelevant to me.

Example: Kate works at the Sweet Shop, a store that sells candy by the pound. One slow day when Kate has nothing else to do, she decides to make a candy mixture of gummy worms and Swedish fish, which ends up costing $3.50 per pound. If she uses 3 pounds of gummy worms costing $4 a pound, how many pounds of Swedish fish costing $2 a pound does she use?

Doesn't the Sweet Shop sell any chocolate? With all this talk about worms and fish, Kate might as well work at a bait-and-tackle shop.

Solution: Set up a chart to organize this information. The chart will be similar to those we used before, except that percent strength is replaced by price. Fill in all the information you are directly given in the question as follows:

Candy	Price/pound	Pounds	Total Price
Worms	$4	3	
Fish	$2		
Mix	$3.50		

Since you're looking for the number of pounds of Swedish fish used in the candy mixture, let that be x. Since the candy mixture is the total number of pounds of both candies, it is $x + 3$. Now that the price and pounds columns are filled in, you can multiply them to find the total price for each row, as shown here.

Candy	Price/pound	Pounds	Total Price
Worms	$4	3	3($4)
Fish	$2	x	$2x$
Mix	$3.50	$x + 3$	$3.50(x + 3)$

Now you can set up an equation to solve for x:

$$3(\$4.00) + \$2.00x = \$3.50(x + 3)$$
$$\$12.00 + \$2.00x = \$3.50x + \$10.50$$
$$\$12.00 = \$1.50x + \$10.50$$
$$\$1.50 = \$1.50x$$
$$1 = x$$

So, Kate needs 1 pound of Swedish fish to make 4 pounds of the mixture. It makes sense that the mixture would include more pounds of gummy worms than Swedish fish, since $4 is closer to $3.50 than $2 is.

Four pounds of candy? She better share that with her friends!

Example: A total of 300 tickets were sold for the high school play, raising a total of $1,250. Advance tickets cost $3 each and tickets sold at the door cost $5 each. How many of each type of ticket were sold?

Even without this book, I can see that it pays to buy early.

Solution: Make another chart and fill in all the information directly given in the question. One difference here is that there isn't really a mixture. The third line of the chart represents the total tickets sold, a special type of ticket with a price between $3 and $5. You'll still solve in the same way, adding the first two rows and setting them equal to the third.

Tickets	Price	Number Sold	Total Money
Advance	$3		
Door	$5		
Total	N/A	300	

Let the number of $3 tickets be x. Since a total of 300 tickets were sold, $300 - x$ is the number of $5 tickets sold. Now that the Price and Number Sold columns are filled in, you can multiply them to find the total price for each row.

Tickets	Price	Number Sold	Total Money
Advance	$3	x	$3x$
Door	$5	$300 - x$	$5(300 - x)$
Total	N/A	300	$1,250

Now, you can set up an equation to solve for x:

$$\$3x + \$5(300 - x) = \$1,250$$
$$\$3x + \$1,500 - \$5x = \$1,250$$
$$-\$2x + \$1,500 = \$1,250$$
$$-\$2x = -\$250$$
$$x = 125$$

So, 125 tickets were sold in advance and $300 - 125 = 175$ were sold at the door.

Get Wise!

Now try the following mixture problems involving prices.

1. A gourmet coffee shop mixes its own blend of coffee beans. How many pounds of beans selling for $7 a pound must be mixed with 9 pounds of beans selling for $12 a pound to make a blend that sells for $10 a pound?

   (A) 6   **(D)** 9
   (B) 7   **(E)** 10
   (C) 8

2. On Saturday, a circus sells 500 admission tickets for a total of $3,900. If an adult ticket costs $12 and a child's ticket costs $5, how many of each type of ticket were sold?

   (A) 300 adult, 200 child
   (B) 275 adult, 225 child
   (C) 250 adult, 250 child
   (D) 225 adult, 275 child
   (E) 200 adult, 300 child

3. How many pounds of pecans selling at $8 a pound must be mixed with 10 pounds of walnuts selling at $5 a pound to make a mixture that sells for $6 a pound?

   (A) 4   **(D)** 7
   (B) 5   **(E)** 8
   (C) 6

4. Cassandra bought 25 yards of fabric at a sale, for a total of $260. She bought two different types of fabric. If Fabric A cost $8 a yard and Fabric B cost $12 a yard, how many yards of each fabric did she buy?

(A) 5 yards of Fabric A, 20 yards of Fabric B

(B) 10 yards of Fabric A, 15 yards of Fabric B

(C) 12.5 yards of Fabric A, 12.5 yards of Fabric B

(D) 15 yards of Fabric A, 10 yards of Fabric B

(E) 20 yards of Fabric A, 5 yards of Fabric B

5. A wholesaler sells tea leaves by the pound. How many pounds of leaves selling for $10 a pound must be mixed with 10 pounds of leaves selling for $7 a pound to make a mixture that sells for $9 a pound?

(A) 5   (D) 20

(B) 10   (E) 30

(C) 15

How Wise?

Check your answers on page 207.

A Word to the Wise

* ★ Don't be intimidated if the question includes a scientific situation. Focus on the math of the situation and use our method to solve.

* ★ Organize the information you are given in a chart that has a row for each solution/item and a column for each characteristic. Use this chart to help you write an equation.

* ★ The percent of the solution multiplied by the amount of the solution gives you the total percent. Likewise, the price of the item multiplied by the number of items gives you the total price.

* ★ Label the unknown value you must solve for with a variable, like x, n, and so on. If you're given the total number of liters/items and neither individual value, label one x and the other as the total minus x.

ANSWERS TO CHAPTER 9: PRACTICE EXERCISES

Mixture Problems (Page 196)

1. **The correct answer is (D).** Use a chart to organize the given information, then set up an equation to solve for x.

Solution	% of Saline	Liters	Total Liters Saline
Solution 1	.30	x	$.30x$
Solution 2	.10	12.5	$.10(12.5)$
Mixture	.25	$x + 12.5$	$.25(x + 12.5)$

$$.30x + .10(12.5) = .25(x + 12.5)$$
$$.30x + 1.25 = .25x + 3.125$$
$$.05x + 1.25 = 3.125$$
$$.05x = 1.875$$
$$x = 37.5$$

2. **The correct answer is (B).** Use a chart to organize the given information, then set up an equation to solve for x.

Paint	% of Pigment	Gallons	Total Gallons Pigment
Paint 1	.10	x	$.10x$
Paint 2	.50	$40 - x$	$.50(40 - x)$
Mixture	.25	40	$.25(40)$

$$.10x + .50(40 - x) = .25(40)$$
$$.10x + 20 - .50x = 10$$
$$-.40x + 20 = 10$$
$$-.40x = -10$$
$$x = 25, \; 40 - x = 15$$

3. **The correct answer is (A).** Use a chart to organize the given information, then set up an equation to solve for x.

Beverage	% of Juice	Liters	Total Liters Juice
Beverage 1	.20	x	.20x
Beverage 2	.50	15	.50(15)
Punch	.35	$x + 15$	.35($x + 15$)

$$.20x + .50(15) = .35(x + 15)$$
$$.20x + 7.5 = .35x + 5.25$$
$$7.5 = .15x + 5.25$$
$$2.25 = .15x$$
$$15 = x$$

4. **The correct answer is (B).** Use a chart to organize the given information, then set up an equation to solve for x.

Solution	% of Acid	Liters	Total Liters Acid
Solution A	.30	18	.30(18)
Solution B	x	6	6x
Mixture	.25	24	.25(24)

$$.30(18) + 6x = .25(24)$$
$$5.4 + 6x = 6$$
$$6x = .6$$
$$x = .10 = 10\%$$

5. **The correct answer is (C).** Use a chart to organize the given information, then set up an equation to solve for x.

Solution	% of Saline	Liters	Total Liters Saline
Solution 1	.80	x	.80x
Solution 2	.10	30	.10(30)
Mixture	.50	$x + 30$	.50($x + 30$)

$$.80x + .10(30) = .50(x + 30)$$
$$.80x + 3 = .50x + 15$$
$$.3x + 3 = 15$$
$$.3x = 12$$
$$x = 40$$

Mixture Problems Involving Prices (Page 202)

1. **The correct answer is (A).** Use a chart to organize the given information, then set up an equation to solve for x.

Coffee	Price/pound	Pounds	Total Price
Coffee 1	$7	x	7x$
Coffee 2	$12	9	$12(9)
Blend	$10	$x + 9$	$10($x + 9$)

$$\$7x + \$12(9) = \$10(x + 9)$$
$$\$7x + \$108 = \$10x + \$90$$
$$\$108 = \$3x + 90$$
$$\$18 = \$3x$$
$$6 = x$$

2. **The correct answer is (E).** Use a chart to organize the given information, then set up an equation to solve for x.

Tickets	Price	Number	Total Price
Adult	$12	x	$12x
Child	$5	$500 - x$	$5(500 - x)$
Total	N/A	500	$3,900

$$\$12x + \$5(500 - x) = \$3,900$$
$$\$12x + \$2,500 - \$5x = \$3,900$$
$$\$7x + \$2,500 = \$3,900$$
$$\$7x = \$1,400$$
$$x = 200, 500 - x = 300$$

3. **The correct answer is (B).** Use a chart to organize the given information, then set up an equation to solve for x.

Nut	Price/pound	Pounds	Total Price
Pecan	$8	x	$8x
Walnut	$5	10	$5(10)
Mix	$6	$x + 10$	$6(x + 10)$

$$\$8x + \$5(10) = \$6(x + 10)$$
$$\$8x + \$50 = \$6x + \$60$$
$$\$8x = \$6x + \$10$$
$$\$2x = \$10$$
$$x = 5$$

4. **The correct answer is (B).** Use a chart to organize the given information, then set up an equation to solve for x.

Fabric	Price/yard	Yards	Total Price
Fabric A	$8	x	$8x
Fabric B	$12	$25 - x$	$12(25 - x)$
Total	N/A	25	$260

$$\$8x + \$12(25 - x) = \$260$$
$$\$8x + \$300 - \$12x = \$260$$
$$-\$4x + \$300 = \$260$$
$$-\$4x = -\$40$$
$$x = 10, 25 - x = 15$$

5. **The correct answer is (D).** Use a chart to organize the given information, then set up an equation to solve for x.

Tea	Price/pound	Pounds	Total Price
Tea 1	$10	x	$10x
Tea 2	$7	10	$7(10)
Mix	$9	$x + 10$	$9(x + 10)$

$$\$10x + \$7(10) = \$9(x + 10)$$
$$\$10x + \$70 = \$9x + \$90$$
$$\$x + \$70 = \$90$$
$$\$x = \$20$$
$$x = 20$$

chapter 10

So Many Variables, So Little Time

UNDERSTANDING MULTIVARIABLE WORD PROBLEMS

A big part of dealing with word problems is figuring out what to do with all the numbers you're given. But what if you're not given numbers, but instead a variable, or two, or three? Don't panic, we'll give you a method that will help you handle these problems with the least amount of pain possible.

A math question with letters instead of numbers? Are we in the twilight zone or something?

If you've done any algebra, you've dealt with variables—the letters that stand in for numbers in math problems. You treat them just as you would numbers, so working with them shouldn't be a big deal, right? Well, the problem is that variables tend to make questions more abstract. When you're dealing with numbers, you kind of know where you stand and can figure out actual values to see if they're reasonable. You can't really get a sense of the answer from just variables. So what do you do? Well, when you have multiple-choice answers, you can make up numbers for the variables.

OK, I'm starting to get used to making up numbers, but that doesn't mean I like it! I think it's time to look into child labor laws! Better yet, they should have a kids' union to protect us from stuff like this.

Example: If p pizzas cost d dollars, how much do $p + 1$ pizzas cost?

(A) $d + 1$

(B) $dp + 1$

(C) $\dfrac{d}{p+1}$

(D) $\dfrac{dp + d}{p}$

(E) $\dfrac{d + 1}{p + 1}$

Textbook algebra method: Set up a proportion relating the cost for p pizzas to the cost of $p + 1$ pizzas. Remember, a proportion is two ratios set equal to each other. Since you know that the price for each individual pizza is the same, you know that two ratios relating the total price to the number of pies must be equal.

Be sure to put the same term in the same place in each proportion. In other words, if you put the price on the top of the first ratio and the number of pizzas on the bottom, you must do the same for the second ratio.

The first ratio is $\dfrac{d}{p}$. Since the price is on top in this ratio, it must be on top in the second ratio. The price is the unknown amount that you're asked to find, so let it be x. This makes your second ratio $\dfrac{x}{p+1}$. Set the two ratios equal so that you have a proportion:

$$\frac{d}{p} = \frac{x}{p+1}$$

Since you want to know the price of $p + 1$ pizzas, cross-multiply and solve for x:

$$d(p+1) = px$$
$$dp + d = px$$
$$\frac{dp+d}{p} = \frac{px}{p}$$
$$\frac{dp+d}{p} = x$$

Making up numbers method: Make up some numbers for the variables. Use numbers that are small and easy to work with. Let's say that $p = 2$, which means that there are two pizzas. Now pick a number for d that's divisible by 2, so your calculations will be simple. Say $d = 10$. Since two pizzas cost 10 dollars, each pizza costs $10 \div 2 = 5$ dollars. You're asked to find the price of $p + 1$ pizzas, or $2 + 1 = 3$ pizzas. Since each pizza costs 5 dollars, the price will be $3(5) = 15$ dollars.

A pizza for 5 bucks? This really must be the twilight zone. Where else could you get a pizza so cheap? I'm sure there are no toppings on this pizza!

OK, so $5 doesn't sound like a realistic price for a pizza. It doesn't matter whether the numbers you make up make sense in the situation. It's much more important that the values are easy to work with. Now that you've come up with the unknown value using the numbers you made up, you have to figure out which answer choice is correct. Do this by plugging in $p = 2$ and $d = 10$ into the answers choices to see which one gives you 15:

(A) $d + 1$: $10 + 1 = 11$ Eliminate!

(B) $dp + 1$: $(10)(2) + 1 = 20 + 1 = 21$ Eliminate!

(C) $\dfrac{d}{p+1}$: $\dfrac{10}{2+1} = \dfrac{10}{3} = 3\dfrac{1}{3}$ Eliminate!

(D) $\dfrac{dp+d}{p}$: $\dfrac{(10 \times 2) + 10}{2} = \dfrac{20+10}{2} = \dfrac{30}{2} = 15$ **Correct!**

(E) $\dfrac{d+1}{p+1}$: $\dfrac{10+1}{2+1} = \dfrac{11}{3} = 3\dfrac{2}{3}$ Eliminate!

WISE NOTE

When you make up numbers for a question with variables, you must plug them into every answer choice. It is possible that more than one choice will match your result. If this happens, try new numbers for just those answer choices until only one matches. If you run out of time, you can guess from the matching choices. To avoid multiple matches, don't pick the number 1, or use the same number for more than one variable.

The by-the-book algebra approach wasn't ridiculously hard, but it can get confusing when you're working with so many variables instead of numbers. In questions with three different variables, it can get pretty tricky. And when you come up with the solution, you really don't have a way to be sure it's correct. When you make up numbers, you know that you have the right answer when you find the matching result. It's also easier for many people to work with numbers instead of variables. This may only be because the variables frighten some people off, but hey, you do what works. Let's try another question, this time with three variables.

Example: A store sells hot cocoa mix for t dollars a tin. Each tin makes p pots of cocoa. If each pot makes m mugs of cocoa, what is the cost of cocoa mix needed to make one mug of cocoa?

(A) tpm (D) $\dfrac{tm}{p}$

(B) $\dfrac{m}{pt}$ (E) $\dfrac{tp}{m}$

(C) $\dfrac{t}{pm}$

Solution: The problem with solving this problem algebraically is that you either get it or you don't. If you're not sure which numbers to multiply and which to divide, you're stuck. So, it's just easier to make up numbers.

Again, make up numbers that are easy to work with. Say that a tin of hot cocoa mix costs 8 dollars, or $t = 8$. Now pick a number for p that divides evenly into 8, so your calculations will be simple. Say $p = 2$. Since an 8-dollar tin makes 2 pots of cocoa, each pot costs $8 \div 2 = 4$ dollars. Now pick a number for m that divides evenly into 4, say $m = 4$. Since each pot of cocoa costs 4 dollars and each pot makes 4 mugs, each mug costs $4 \div 4 = 1$ dollar.

Now that you've come up with the unknown value using the numbers you made up, you have to figure out which answer choice is correct. Do this by plugging $t = 8$, $p = 2$, and $m = 4$ into the given answer choices to see which one gives you 1:

(A) tpm: $(8)(2)(4) = 64$ Eliminate!

(B) $\dfrac{m}{pt}$: $\dfrac{4}{(2)(8)} = \dfrac{4}{16} = \dfrac{1}{4}$ Eliminate!

(C) $\dfrac{t}{pm}$: $\dfrac{8}{(2)(4)} = \dfrac{8}{8} = 1$ **Correct!**

(D) $\dfrac{tm}{p}$: $\dfrac{(8)(4)}{2} = \dfrac{32}{2} = 16$ Eliminate!

(E) $\dfrac{tp}{m}$: $\dfrac{(8)(2)}{4} = \dfrac{16}{4} = 4$ Eliminate!

When do we eat? All this talk about pizza and hot cocoa is making me hungry. I want _p_ slices of pizza and after that _m_ mugs of cocoa for _c_ = 0 dollars!

Get Wise!

Now for some variety. Try the following multivariable word problems.

1. A candy store owner buys molasses chews in bulk and then divides them into cellophane bags to sell to his customers. He has m pieces of molasses chews and plans to divide them evenly among b bags. The bags are smaller than he thought, so he needs to use an additional x bags. How many molasses chews will each bag contain?

(A) $\dfrac{m}{x}$

(D) $\dfrac{b+x}{m}$

(B) $\dfrac{m}{b}+x$

(E) $\dfrac{m}{b+x}$

(C) $\dfrac{x}{b+m}$

I've got an idea. How about I run over to the candy store and buy a bag of molasses chews? Then we can just count the number of pieces instead of messing with all these variables.

2. Peter can mow g square feet of grass in h hours. How many square feet of grass can he mow in $h-1$ hours?

(A) $g-1$

(D) $\dfrac{h^2-h}{g}$

(B) $\dfrac{g-1}{h}$

(E) $\dfrac{g}{h-1}$

(C) $\dfrac{gh-g}{h}$

3. If b baseballs weigh p pounds, how much do n baseballs weigh?

(A) $p + n$ **(D)** $\dfrac{p}{n+b}$

(B) $\dfrac{np}{b}$ **(E)** $\dfrac{b}{n+p}$

(C) $\dfrac{bn}{p}$

4. Brenna writes w pages of her report in d days. How many days does it take her to write $w + 10$ pages?

(A) $d + 10$ **(D)** $\dfrac{dw+10w}{d}$

(B) $10d$ **(E)** $\dfrac{w+10}{d}$

(C) $\dfrac{dw+10d}{w}$

5. In a board game, r red pieces are worth p points. How many points are $r + 5$ red pieces worth?

(A) $5p$ **(D)** $\dfrac{pr+5p}{r}$

(B) $\dfrac{5p}{r}$ **(E)** $\dfrac{pr+r}{5p}$

(C) $\dfrac{r+5}{p}$

How Wise?

Check your answers on page 225.

MULTIVARIABLE WORD PROBLEMS WITH DIFFERENT UNITS

Some multivariable word problems introduce another wrinkle. They express a rate in two different units, like hours and minutes, days and weeks, pounds and ounces, and so on. This means one of the units needs to be converted, using a unit conversion factor. In a question involving hours and minutes, for example, the conversion is 1 hour = 60 minutes. This means that the correct expression will include this number. But where does it go? It's just one more piece of information for you to wrestle with. Making up numbers is especially helpful to help you tackle these problems.

So it's like doing extra math inside a math problem. That's too much!

Example: Emeril buys fresh pasta at the market, paying *d* dollars for *o* ounces. How much would he have to pay for *p* pounds of this pasta?

(A) $\dfrac{dp}{o}$ (D) $\dfrac{16o}{dp}$

(B) $\dfrac{dp}{16o}$ (E) $\dfrac{o}{16dp}$

(C) $\dfrac{16dp}{o}$

Pasta—here we go again with food. I should have done this chapter after lunch. If you don't see me for a few pages, it's because I'm off somewhere grabbing a snack.

Solution: Make up numbers for the variables that are easy to work with. Say that the pasta costs 10 dollars, or $d = 10$. Pick a number for ounces, o, that divides evenly into 10, say $o = 5$. Since Emeril paid 10 dollars for 5 ounces of pasta, he paid $10 \div 5 = 2$ dollars an ounce. Now you want to know how much p pounds will cost, so pick a number for p. Say $p = 2$.

You don't have a per-pound rate, just a per-ounce rate. Here's where the conversion factor comes in. Since there are 16 ounces in a pound, there are $16(2) = 32$ ounces in 2 pounds. So 2 pounds of pasta is the same as 32 ounces of pasta, which costs $32(2) = 64$ dollars.

With pasta costing that much, it's no wonder that Emeril couldn't afford to buy any sauce!

Now that you've come up with the unknown value using the numbers you made up, you have to figure out which answer choice is correct. Do this by plugging in $d = 10$, $o = 5$, and $p = 2$ into the answer choices to see which one gives you 64:

(A) $\dfrac{dp}{o}$: $\dfrac{(10)(2)}{5} = \dfrac{20}{5} = 4$ Eliminate!

(B) $\dfrac{dp}{16o}$: $\dfrac{(10)(2)}{16(5)} = \dfrac{20}{80} = \dfrac{1}{4}$ Eliminate!

(C) $\dfrac{16dp}{o}$: $\dfrac{16(10)(2)}{5} = \dfrac{320}{5} = 64$ **Correct!**

(D) $\dfrac{16o}{dp}$: $\dfrac{16(5)}{(10)(2)} = \dfrac{80}{20} = 4$ Eliminate!

(E) $\dfrac{o}{16dp}$: $\dfrac{5}{16(10)(2)} = \dfrac{5}{320} = \dfrac{1}{64}$ Eliminate!

WISE NOTE

You do have to plug your made-up numbers into every choice, but you don't have to figure out their exact values. If you can see that the result is going to be smaller or larger than the number you are looking for, you can eliminate it. For example, in the problem we just looked at, you knew you were looking for a whole number. When you plug your made-up numbers into answer choice (E), it's pretty clear that the result will be a fraction less than 1. Since you can tell that it's not going to be the value you're looking for, you can eliminate it without spending the time multiplying and dividing to find its exact value.

Get Wise!

Now work out the following multivariable word problems with different units.

1. Miranda jogs a miles every b hours. At this rate, how many miles will Miranda jog in c minutes?

(A) $\dfrac{ab}{60c}$

(D) $\dfrac{60ac}{b}$

(B) $\dfrac{ac}{60b}$

(E) $\dfrac{60a}{bc}$

(C) $\dfrac{60ab}{c}$

2. Spencer puts a miles on his car every b weeks. At this rate, how many miles will he put on his car in c days?

(A) $\dfrac{ac}{7b}$

(D) $\dfrac{7ab}{c}$

(B) $\dfrac{ab}{7c}$

(E) $\dfrac{abc}{7}$

(C) $\dfrac{7ac}{b}$

3. Sierra decorates c chocolate chip cookies in m minutes. At this rate, how many cookies will she decorate in h hours?

(A) $\dfrac{60ch}{m}$

(D) $\dfrac{60h}{cm}$

(B) $\dfrac{ch}{60m}$

(E) $\dfrac{chm}{60}$

(C) $\dfrac{cm}{60h}$

Finally we're talking desserts! Pizza and pasta are good, but nothing's better than a chocolate chip cookie!

4. Federal Parcel Service charges a shipping fee of *d* dollars for every *p* pounds. At this rate, how much would it cost to ship a package weighing *o* ounces?

    (A) $16dop$

    (B) $\dfrac{16do}{p}$

    (C) $\dfrac{do}{16p}$

    (D) $\dfrac{16p}{do}$

    (E) $\dfrac{dp}{16o}$

5. Nigella makes *b* bagels in *h* hours. At this rate, how long will it take her to make *d* dozen bagels?

    (A) $12h$

    (B) $\dfrac{12bd}{h}$

    (C) $\dfrac{12dh}{b}$

    (D) $\dfrac{12d}{bh}$

    (E) $\dfrac{12h}{bd}$

How Wise?

Check your answers on page 226.

A Word to the Wise

★ When a multivariable word problem has answer choices involving variables, you can make up numbers for the variables. Solve the problem with these numbers. Then plug these numbers into the answer choices to see which matches.

★ You must plug your numbers into every answer choice. If more than one match, try new numbers in those answer choices only, until just one matches.

★ To avoid multiple matching answer choices, avoid using the number 1 and plugging in the same number for more than one variable.

★ Making up numbers is especially helpful when multivariable questions use two different measuring units.

★ You do have to plug your numbers into every answer choice, but you don't have to find their exact values if you can tell that they will be bigger or smaller than the value you're looking for.

ANSWERS TO CHAPTER 10: PRACTICE EXERCISES

Multivariable Word Problems (Page 217)

1. **The correct answer is (E).** Make up numbers for the variables that are easy to work with. Pick a number for m, say 12. These candies are to be divided evenly into b bags, so pick a number that divides evenly into 12, say $b = 4$. These bags, plus x more, will ultimately be used to divide up the candy. Pick a value for x that when added to 4 will divide evenly into 12. Say $x = 2$. This means that 12 pieces of candy will be divided among $4 + 2 = 6$ bags, so each bag will hold $12 \div 6 = 2$ pieces of candy. If you plug $m = 12$, $b = 4$, and $c = 2$ into the answer choices, only answer choice (E) gives you 2:

$$\frac{m}{b+x} : \quad \frac{12}{4+2} = \frac{12}{6} = 2$$

2. **The correct answer is (C).** Make up numbers for the variables that are easy to work with. Say that $g = 10$. Since Peter can mow 10 square feet of lawn in h hours, pick a number for h that divides into 10. Say $h = 5$. This means that Peter mows $10 \div 5 = 2$ square feet of lawn in an hour. So in $h - 1 = 4$ hours, he mows $2(4) = 8$ square feet of lawn. If you plug $g = 10$ and $h = 5$ into the answer choices, only choice (C) gives you 8:

$$\frac{gh - g}{h} : \quad \frac{(10)(5) - 10}{5} = \frac{50 - 10}{5} = \frac{40}{5} = 8$$

3. **The correct answer is (B).** Make up numbers for the variables that are easy to work with. If $b = 2$ and $p = 8$, then each baseball weighs $8 \div 2 = 4$ pounds. If $n = 3$, then 3 baseballs weigh $3(4) = 12$ pounds. If you plug $b = 2$, $p = 8$, and $n = 3$ into the answer choices, only choice (B) gives you 12:

$$\frac{np}{b} : \quad \frac{(3)(8)}{2} = \frac{24}{2} = 12$$

4. **The correct answer is (C).** Make up numbers for the variables that are easy to work with. If $w = 10$ and $d = 5$, then Brenna writes $10 \div 5 = 2$ pages each day. So writing $w + 10 = 20$ pages would take her $20 \div 2 = 10$ days. If you plug $w = 10$ and $d = 5$ into the answer choices, only choice (C) gives you 10:

$$\frac{dw + 10d}{w} : \frac{(5)(10) + 10(5)}{10} = \frac{50 + 50}{10} = \frac{100}{10} = 10$$

5. **The correct answer is (D).** Make up numbers for the variables that are easy to work with. If $r = 2$ and $p = 10$, then each red piece is worth $10 \div 2 = 5$ points. So $r + 5 = 7$ pieces are worth $7(5) = 35$ points. If you plug $r = 2$ and $p = 10$ into the answer choices, only choice (D) gives you 35:

$$\frac{pr + 5p}{r} : \frac{(10)(2) + 5(10)}{2} = \frac{20 + 50}{2} = \frac{70}{2} = 35$$

Multivariable Word Problems with Different Units (Page 222)

1. **The correct answer is (B).** Make up numbers for the variables that are easy to work with. If $a = 120$ and $b = 2$, then Miranda jogs $120 \div 2 = 60$ miles each hour. Since there are 60 minutes in an hour, this means that Miranda jogs 1 mile every minute. If $c = 30$, then Miranda jogs $1(30) = 30$ miles in 30 minutes. If you plug $a = 120$, $b = 2$, and $c = 30$ into the answer choices, only choice (B) gives you 30:

$$\frac{ac}{60b} : \frac{(120)(30)}{60(2)} = \frac{3,600}{120} = 30$$

2. **The correct answer is (A).** Make up numbers for the variables that are easy to work with. If $a = 21$ and $b = 3$, Spencer puts $21 \div 3 = 7$ miles on his car each week. Since there are 7 days in a week, Spencer puts $7 \div 7 = 1$ mile on his car every day. If $c = 5$, then Spencer puts $5(1) = 5$ miles on his car in 5 days. If you plug $a = 21$, $b = 3$, and $c = 5$ into the answer choices, only choice (A) gives you 5:

$$\frac{ac}{7b} : \frac{(21)(5)}{7(3)} = \frac{(21)(5)}{21} = 5$$

3. **The correct answer is (A).** Make up numbers for the variables that are easy to work with. If $c = 30$ and $m = 15$, then Sierra decorates $30 \div 15 = 2$ chocolate chip cookies in one minute. Since there are 60 minutes in an hour, this means that Sierra decorates $2(60) = 120$ cookies each hour. If $h = 2$, then Sierra decorates $2(120) = 240$ cookies in 2 hours. If you plug $c = 30$, $m = 15$, and $h = 2$ into the answer choices, only choice (A) gives you 240:

$$\frac{60ch}{m} : \frac{60(30)(2)}{15} = 60(2)(2) = 240$$

4. **The correct answer is (C).** Make up numbers for the variables that are easy to work with. If $d = 32$ and $p = 2$, then the shipping charge is $32 \div 2 = 16$ dollars per pound. Since there are 16 ounces in a pound, the charge is $16 \div 16 = 1$ dollar per ounce. If $o = 5$, then it costs $1(5) = 5$ dollars to ship a 5-ounce package. If you plug $d = 32$, $p = 2$, and $o = 5$ into the answer choices, only choice (C) gives you 5:

$$\frac{do}{16p} : \frac{(32)(5)}{16(2)} = \frac{(32)(5)}{32} = 5$$

5. **The correct answer is (C).** Make up numbers for the variables that are easy to work with. If $b = 24$ and $h = 2$, then Nigella makes $24 \div 2 = 12$ bagels each hour. There are 12 bagels in a dozen, so this means that Nigella makes 1 dozen bagels each hour. If $d = 3$, then it takes Nigella $1(3) = 3$ hours to make 3 dozen bagels. If you plug $b = 24$, $h = 2$, and $d = 3$ into the answer choices, only choice (C) gives you 3:

$$\frac{12dh}{b} : \frac{12(3)(2)}{24} = \frac{(3)(24)}{24} = 3$$

chapter 11

What Are the Odds?

UNDERSTANDING PROBABILITY

Have you ever daydreamed about winning the lottery? About what you'd do with the oodles of money, about which friends you would treat to a ride in your new chauffeured limousine, about how your chance of winning is about $\dfrac{1}{32,468,436}$? When you talk about the likelihood of an event occurring, you're talking about probability.

I may not know much about probability yet, but even I know that those chances are pretty lame.

The probability of any event occurring is between 0 and 1. The closer the probability is to 1, the more likely it is that it will happen. If the probability is 0, there is no chance that it will occur. If the probability is 1, it is certain that the event will occur. So looking again at the probability of you winning the lottery, we'd advise you not to get your hopes up.

The mathematical formula for probability is:

$$\text{Probability} = \frac{\text{Outcomes you want}}{\text{Total possible outcomes}}$$

For example, let's say that your parents want you or your brother to wash the dishes. What is the probability that you'll have to do it? Since either you or your brother could do it, there are two possible outcomes. Since there's one of you, there is one outcome you want. (Well, in this case it's not the outcome that you want exactly, but it's the outcome that fits the condition.) That makes the probability $\frac{1}{2}$ that you'll get dish duty.

I don't care what the probability formula says. I'm certain that I'd have to be the one scrubbing away. My spoiled little brother never has to do anything.

A word problem might ask you to find the probability of something NOT happening. That's easy enough to do. The probability of an event NOT occurring plus the probability of that same event occurring is always equal to 1. So if your math teacher tells you that the probability that he'll give you a quiz tomorrow is $\dfrac{1}{4}$, then the probability that he will NOT give you a quiz tomorrow is $1 - \dfrac{1}{4} = \dfrac{3}{4}$.

Finally, the odds are in my favor. I probably won't win the lottery and I may get stuck with dishpan hands, but at least it seems likely that I won't have to suffer through a math quiz tomorrow.

WISE NOTE

Probability is usually written as a fraction, but it can also be written as a decimal or a percent. For example, if there is a 1-in-4 chance of something happening, the probability that it will happen is:

$$\frac{1}{4} = .25 = 25\%.$$

There are certain situations that come up a lot in probability word problems. Some involve marbles of different colors in a bag, different colored socks in a drawer, flipping a coin, rolling a die, spinning a spinner, or picking from a deck of playing cards. The questions about a coin, die (which is the singular of dice), or deck of cards may specify that they are "fair." This means that the coin has two sides, a heads and tails, each of which is just as likely to come up as the other. The die is a cube with one of the values 1 through 6 marked on its sides, any of which is just as likely to come up as the others. There are 52 cards in a standard deck, 13 in each of four different suits: hearts, diamonds, spades, and clubs.

Example: Greg and Sophie are playing a board game. It's Greg's turn to roll the die, and if he rolls a 3 or higher, he wins the game. What is the probability that Greg will win the game? (The die is a fair six-sided die.)

Solution: Plug these values into the probability formula. The die has a total of six sides with values 1 through 6. Greg could roll any one of them, so the total possible outcome is 6. You're asked to find the probability that Greg will win. He wins if he rolls a value of 3 or higher, so he will win if he rolls a 3, 4, 5, or 6. That means that there are 4 outcomes that you want. So the probability that Greg will win is:

$$\text{Probability} = \frac{\text{Outcomes you want}}{\text{Total possible outcomes}}$$

$$\text{Probability} = \frac{4}{6}$$

$$\text{Probability} = \frac{2}{3}$$

WISE POINT

Keep it simple: Always reduce a fractional probability to its simplest form.

Example: Monica has one extra concert ticket. She is trying to decide whether to take Phoebe, Rachel, Joey, Ross, or Chandler with her. What is the probability that she will NOT take one of the girls to the concert with her?

Solution: Find the probability that Monica will take one of the girls to the concert with her. Then subtract that fraction from 1 to find the probability that she will NOT take one of the girls with her.

Since there are 5 possibilities for whom Monica will take with her to the concert, the total possible outcomes is 5. Of those 5, 2 are girls: Phoebe and Rachel. So the probability that Monica will take one of the girls with her is:

$$\text{Probability} = \frac{\text{Outcomes you want}}{\text{Total possible outcomes}}$$

$$\text{Probability} = \frac{2}{5}$$

So the probability that Monica will NOT take one of the girls with her is

$$1 - \frac{2}{5} = \frac{3}{5}$$

Get Wise!

This isn't so hard, right? Test yourself with the following probability problems.

1. There are 24 students in the class and 16 of them are boys. If a student is selected at random, what is the probability that the student is a girl?

(A) $\dfrac{1}{24}$

(D) $\dfrac{1}{2}$

(B) $\dfrac{1}{4}$

(E) $\dfrac{2}{3}$

(C) $\dfrac{1}{3}$

2. Holden sold 45 tickets for the school raffle. If a total of 500 tickets were sold, what is the probability that Holden sold the winning ticket?

(A) $\dfrac{1}{45}$

(D) $\dfrac{9}{10}$

(B) $\dfrac{9}{100}$

(E) $\dfrac{91}{100}$

(C) $\dfrac{9}{50}$

3. David selects a card at random from a standard deck of playing cards. What is the probability that this card is NOT a heart? (A standard deck contains 52 cards, which are divided equally into four suits: hearts, diamonds, spades, and clubs.)

(A) $\dfrac{1}{52}$

(D) $\dfrac{3}{4}$

(B) $\dfrac{1}{13}$

(E) $\dfrac{12}{13}$

(C) $\dfrac{1}{4}$

4. Izzy's track team consists of 3 sophomores, 15 juniors, and 6 seniors. If one student is selected to serve as captain, what is the probability that the captain will be a senior?

(A) $\dfrac{1}{7}$ (D) $\dfrac{2}{3}$

(B) $\dfrac{1}{4}$ (E) $\dfrac{3}{4}$

(C) $\dfrac{1}{3}$

5. Carol isn't tall enough to be able to read the titles of the 5 textbooks she has on the top shelf of her locker. She needs to bring her science book home to study tonight. If she grabs a book from this shelf, what is the probability that it is NOT her science book?

(A) $\dfrac{1}{5}$ (D) $\dfrac{4}{5}$

(B) $\dfrac{2}{5}$ (E) 1

(C) $\dfrac{3}{5}$

How Wise?

Check your answers on page 250.

COMPLEX PROBABILITY

The problems we've looked at so far have dealt with simple probability. Some questions are a little more complicated. They ask about the probability of one event taking place after another has already happened, or the probability of a couple of events taking place in a row. Say, for instance, you bought a box of assorted chocolates. The chocolates all look alike, so you can't tell what kind of filling each has until you bite into it. Your favorites are the caramels, and you know that there are a few of them in the box. What is the probability that you'll pick out a caramel at random?

I like when the box of chocolates has a map showing you which candy is where. Otherwise, you never know if you're going to bite into some icky candy, like lemon chiffon or some old-fashioned candy like that. That's when you have to be sneaky and poke the chocolates from underneath so you can put back the ones you don't want.

Example: You've just opened your box of 25 assorted chocolates. You read the box and see that it has 10 caramels, your favorite candy. Before you can stop him, your little brother swipes a piece of chocolate that turns out to be a caramel. If now you take a chocolate from the box, what is the probability that it will be a caramel?

Solution: Figure out if the event that has already happened changes the number of outcomes you want and/or the total number of outcomes.

The box starts out with 25 chocolates, 10 of which are caramels. After your brother takes a caramel, there are now just 9 of them left. That's the outcome you want. The total number of chocolates in the box also decreases by 1, making it 24. That's the total number of possible outcomes. So after your brother swipes a caramel, the probability that you'll pick a caramel as well is:

$$\text{Probability} = \frac{\text{Outcomes you want}}{\text{Total possible outcomes}}$$

$$\text{Probability} = \frac{9}{24}$$

$$\text{Probability} = \frac{3}{8}$$

WISE NOTE

It's important to remember to check if the first event changes *both* totals. When you're taking a multiple-choice test, you can bet that some of the wrong answer choices for a question like this would be the result of making the change in one place only. For this

question, these wrong answers would be $\frac{9}{25}$

(forgetting to subtract 1 from the total

outcomes) or $\frac{10}{24} = \frac{5}{12}$ (forgetting to subtract

1 from the outcomes you want).

Some questions ask about the probability of a number of events happening in a row. Say you're playing a game and want to see what the probability is that you'll roll a 5 twice in a row. You do this by figuring out each individual probability and then multiplying them together. In this case, you have a 1-in-6 chance of rolling a 5 on each roll, so the probability of rolling a 5 twice in a row is

$\frac{1}{6} \times \frac{1}{6} = \frac{1}{36}$. When you're rolling a die or flipping a coin, the first event doesn't affect the outcomes for the second. As you've seen in our caramel candy example, though, some events do affect future outcomes. It's important to check if this is the case when you figure out the individual probabilities.

Example: You've bought another box of chocolates. This one has a total of 21 pieces, 9 of which are caramel. After checking to make sure that your brother is nowhere in sight, you open the box of candy. What is the probability that you will select 2 caramels in a row?

Solution: The box starts out with 21 chocolates, 9 of which are caramels. That means that you have a 9-in-21 chance of getting a caramel on the first try. Once you remove a caramel, there are only 9 – 1 = 8 caramels left, and only 21 – 1 = 20 candies total left. That means that you have an 8-in-20 chance of getting a caramel on your second try. So the probability that you'll pick out 2 caramels in a row is:

$$\text{Probability} = \frac{9}{21} \times \frac{8}{20}$$

$$\text{Probability} = \frac{\overset{3}{\cancel{9}}}{\underset{7}{\cancel{21}}} \times \frac{\overset{2}{\cancel{8}}}{\underset{5}{\cancel{20}}}$$

$$\text{Probability} = \frac{6}{35}$$

Get Wise!

Now's your chance to prove that you understand complex probability problems.

1. Sandy's sock drawer contains 10 black and 12 white socks. Sandy removes a black sock from the drawer and puts it on. Without looking, she removes another sock from the drawer and puts it on. What is the probability that the second sock Sandy drew matches the first one?

    (A) $\dfrac{5}{11}$ **(D)** $\dfrac{10}{21}$

    (B) $\dfrac{9}{22}$ **(E)** $\dfrac{1}{2}$

    (C) $\dfrac{3}{7}$

I'm not saying that we all have to be slaves to fashion, but is it too much to ask that people look at their clothes before putting them on? Come on!

2. There is a box of assorted flavored ice pops in the freezer. Six are orange, 5 are cherry, and 4 are grape. If 2 pops are removed in a row, what is the probability that both are cherry?

    (A) $\dfrac{4}{45}$ **(D)** $\dfrac{1}{3}$

    (B) $\dfrac{2}{21}$ **(E)** $\dfrac{2}{7}$

    (C) $\dfrac{5}{42}$

3. What is the probability of flipping tails three times in a row with a fair two-sided coin?

(A) $\dfrac{1}{8}$ (D) 1

(B) $\dfrac{1}{4}$ (E) $\dfrac{3}{2}$

(C) $\dfrac{1}{2}$

4. A bag contains 8 red, 7 blue, and 6 green marbles. After a blue marble is removed from the bag, what is the probability that the next marble drawn from the bag will NOT be blue?

(A) $\dfrac{3}{10}$ (D) $\dfrac{2}{3}$

(B) $\dfrac{1}{3}$ (E) $\dfrac{7}{10}$

(C) $\dfrac{1}{2}$

5. Mia removes a 3 from a standard deck of playing cards. What is the probability that the next card she draws will also be a 3?

(A) $\dfrac{1}{52}$ (D) $\dfrac{3}{4}$

(B) $\dfrac{1}{17}$ (E) $\dfrac{16}{17}$

(C) $\dfrac{1}{3}$

How Wise?

Check your answers on page 252.

COMBINATIONS

Sometimes, life gives you lots of choices. When you have a closet full of clothes, you have to spend some time deciding what to wear to school each day. Did you ever stop to think how many different outfits you could put together from your wardrobe? If you did, you've done some thinking about combinations.

This is my kind of math. Now the next time my mother tells me to hurry up and decide what to wear to school, I can tell her I'm doing mathematical research for a combinations question.

In combinations questions, you have to figure out the number of different possibilities for a group of items or circumstances. This could be figuring out how many outfits you can put together or the number of meal combinations that are possible at a salad bar. To find the number of combinations, you use the fundamental counting principle.

Don't be alarmed by that technical terminology. Basically, it means that you multiply together the numbers of choices to find the number of different combinations that are possible. Or, in other words, it means that if there are x ways that one thing can happen, and y ways that another can happen, then there are xy ways that the two things can happen together. You've actually already used this principle earlier in the chapter. You used it when you multiplied together the individual probabilities for two events to find the probability that they would happen in a row.

That's pretty clever, sneaking in that fundamental thingamajig when I wasn't looking. It's like when you're at the doctor's office worried about getting a shot, and the next thing you know it's all over already.

Example: Jamie packs 3 pairs of pants, 4 shirts, and 2 pairs of shoes for his vacation. If an outfit consists of 1 pair of pants, 1 shirt, and 1 pair of shoes, how many different outfits can Jamie make with these clothes?

Solution: Multiply together the number of items for each option. To find the number of different outfits that Jamie can put together, multiply the number of choices for each option. In this case, that means multiplying together the number of pants, the number of shirts, and the number of pairs of shoes. So Jamie can put together $3 \times 4 \times 2 = 24$ different outfits from these clothes.

WISE POINT

Order doesn't matter in a combinations problem. That means that a pair of jeans, a red T-shirt, and a pair of sneakers is the same outfit as a pair of sneakers, a pair of jeans, and a red T-shirt.

Get Wise! Mastering Math Word Problems

Get Wise!

Try some combinations problems on your own.

1. The Burger Hut offers a lunch special that includes 1 sandwich, 1 side dish, and 1 soft drink. If there are 6 sandwich choices, 2 side-dish choices, and 5 soft-drink choices, how many different lunch specials are possible?

(A) 13   (D) 120

(B) 30   (E) 360

(C) 60

That's a lot of combinations that will give you oily skin and clog your arteries. I stick to my combo instead—a large salad, a bottle of water, and a yogurt for dessert.

2. A clothing manufacturer is preparing its fall line. It offers 4 different styles of sweaters, each of which is available in 5 colors and 4 sizes. How many different sweaters will the company produce?

(A) 16   (D) 100

(B) 20   (E) 400

(C) 80

3. A 3-digit number is created using a numeral from Set A in the 100's place, a numeral from Set B in the 10's place, and a numeral from Set C in the 1's place. How many different numbers are possible?

Set A {1, 2, 3}

Set B {4, 5, 6, 7, 8}

Set C {9, 0}

(A) 10

(B) 27

(C) 30

(D) 50

(E) 100

4. A basic pasta dish is made up of 1 type of pasta, 1 type of sauce, and 1 type of topping. If the restaurant offers 10 different pastas, 5 different sauces, and 6 different toppings, how many different basic pasta dishes are possible?

(A) 21

(B) 30

(C) 50

(D) 60

(E) 300

5. A factory produces 10 different models of cars, which are available in 6 exterior colors and 2 interior finishes. How many different cars does this factory produce?

(A) 18

(B) 30

(C) 60

(D) 120

(E) 360

How Wise?

Check your answers on page 254.

PERMUTATIONS

Combinations are arrangements of items where order doesn't matter. Permutations are a similar kind of problem, but the difference is that, with them, order matters. For example, say that you wanted to figure out the total number of different passwords possible for a computer log-in. Clearly the password ABCD is different from the password DACB, even though both use the same 4 letters. In some cases, the number of options involved in a permutations problem makes it easy to write out all the different possibilities. Even in these cases, though, it's easy to make a careless mistake or overlook some choices. And when the number of options increases, it becomes impractical to list out all the possibilities. Lucky for you, we've got a method to help you figure it out.

Example: The password for a computer log-in must be 5 characters long. It is made up of the digits 0 through 9, none of which can be used more than once. How many different passwords are possible?

Solution: This is a permutations question, because the order of digits matters: 12345 is different from 54321.

Sketch a slot for each position. There are 5 digits in the code, so you need 5 slots:

— — — — —

Next, plug in the number of possibilities for each slot. The digits 0 through 9 can each be used only once per code. Be careful—there are 10 digits from 0 through 9: 0, 1, 2, 3, 4, 5, 6, 7, 8, and 9. This means that there are 10 possibilities for the first slot. Since each digit can be used only once, this leaves $10 - 1 = 9$ possibilities for the second slot, $9 - 1 = 8$ for the third slot, $8 - 1 = 7$ for the fourth slot, and $7 - 1 = 6$ for the fifth slot.

$$\underline{10} \quad \underline{9} \quad \underline{8} \quad \underline{7} \quad \underline{6}$$

Multiply them together to find the total number of codes possible:

$$10 \times 9 \times 8 \times 7 \times 6 = 30{,}240$$

Wow, it'd take me 30,240 years to write out all those codes by hand. I like this way better.

Sometimes, permutations questions involve special cases. Say that 4 friends go to the movies and find the last 4 seats in the last row available. If they want to figure out how many different ways they can arrange themselves in these seats, it's a straight permutations question. But what if two friends want to sit next to each other or one person insists on sitting in the aisle seat?

Example: Amy, Bette, Catherine, and Daisy go to the movies together. They find 4 empty seats on the left-hand aisle. If Bette insists on sitting in the aisle seat, how many different ways can these friends be seated?

Solution: Fill in the special condition, and then see how many possibilities are left for the rest of the slots.

There are 4 seats, so set up 4 slots:

__ __ __ __

Bette must sit in the aisle seat, so put her there first:

<u>B</u> _ _ _

Once Bette is seated, Amy, Catherine, and Daisy remain unseated. That means that there are 3 choices for the second seat. Once one of them is seated in the second seat, there are only 2 choices for the third seat, and just 1 choice for the fourth seat.

<u>B</u> <u>3</u> <u>2</u> <u>1</u>

That means that there are $3 \times 2 \times 1 = 6$ ways that the friends can seat themselves if Bette insists on sitting in the aisle seat.

I don't blame Bette for wanting to sit in the aisle seat. That way it's easy to get up during the movie if it's boring, or if you run out of munchies.

Get Wise!

Got it? Great, then you're ready to try the following permutations problems.

1. Eight students are running for the 3 available student council positions. If no student is able to hold more than one office, how many different arrangements of these students in these three positions are possible?

(A) 6   (D) 56

(B) 24   (E) 336

(C) 42

2. In how many different ways can 5 people arrange themselves in a row of 5 seats, if 2 people must sit in end seats?

(A) 6   (D) 25

(B) 12   (E) 120

(C) 24

3. A bank issues alphabetical codes to its customers. The codes are 4 characters long and are created from the 26 letters. If each letter can be used no more than once per code, how many different codes are possible?

(A) 26   (D) $26 \times 25 \times 24 \times 23$

(B) 4×26 (E) $26 \times 26 \times 26 \times 26$

(C) 26×25

4. How many different 3-digit numbers can be created from the digits 2, 4, 6, and 8, if each digit can appear only once per number?

(A) 3   (D) 12

(B) 4   (E) 24

(C) 6

5. In how many different ways can 4 people arrange themselves in a row of 5 seats?

  (A) 20   **(D)** 96

  (B) 24   **(E)** 120

  (C) 48

How Wise?

Check your answers on page 254.

A Word to the Wise

★ Probability is the likelihood that some event will occur and is found

  with this formula: $\text{Probability} = \dfrac{\text{Outcomes you want}}{\text{Total possible outcomes}}$

★ The probability of an event NOT occurring plus the probability of that same event occurring is equal to 1.

★ Always reduce a fractional probability to its simplest form.

★ Find the probability of events happening in a row by multiplying together their individual probabilities.

★ When dealing with complex probabilities, always check to see whether one event's occurrence affects the outcomes for the second event.

★ To figure out the number of possible combinations, multiply together the number of choices for each option. Order doesn't matter in combinations.

★ Order does matter in permutations. To find the number of permutations, sketch a slot for each option, plug the number of possibilities into each slot, and multiply them together.

ANSWERS TO CHAPTER 11: PRACTICE EXERCISES

Probability Problems (Page 234)

1. **The correct answer is (C).** Since 16 of the 24 students are boys, $21 - 16 = 8$ of them are girls. So the number of outcomes you want is 8, and the total number of possible outcomes is 24. That makes the probability of a girl being selected:

$$\text{Probability} = \frac{\text{Outcomes you want}}{\text{Total possible outcomes}}$$

$$\text{Probability} = \frac{8}{24}$$

$$\text{Probability} = \frac{1}{3}$$

2. **The correct answer is (B).** Holden sold 45 tickets, so 45 is the number of outcomes you want. Since 500 tickets were sold, that is the total number of possible outcomes. That makes the probability that Holden sold the wining ticket:

$$\text{Probability} = \frac{\text{Outcomes you want}}{\text{Total possible outcomes}}$$

$$\text{Probability} = \frac{45}{500}$$

$$\text{Probability} = \frac{9}{100}$$

3. **The correct answer is (D).** There are 13 cards of each suit in a deck of cards. Since 13 of the cards are hearts, $52 - 13 = 39$ cards are NOT hearts. That makes the outcome you want 39. Since there are 52 cards in the deck, that is the total number of possible outcomes. So the probability of NOT picking a heart is:

$$\text{Probability} = \frac{\text{Outcomes you want}}{\text{Total possible outcomes}}$$

$$\text{Probability} = \frac{39}{52}$$

$$\text{Probability} = \frac{3}{4}$$

4. **The correct answer is (B).** You're asked for the probability that a senior will be captain. There are 6 seniors on the team, so that is the number of outcomes you want. There are a total of $3 + 15 + 6 = 24$ students on the team, so 24 is the total number of possible outcomes. So the probability that a senior will be captain is:

$$\text{Probability} = \frac{\text{Outcomes you want}}{\text{Total possible outcomes}}$$

$$\text{Probability} = \frac{6}{24}$$

$$\text{Probability} = \frac{1}{4}$$

5. **The correct answer is (D).** There are 5 textbooks on the top shelf, so that is the total number of possible outcomes. Since one of them is a science book, $5 - 1 = 4$ of them are NOT the science book. So the outcomes you want is 4. That makes the probability:

$$\text{Probability} = \frac{\text{Outcomes you want}}{\text{Total possible outcomes}}$$

$$\text{Probability} = \frac{4}{5}$$

Complex Probability Problems (Page 239)

1. **The correct answer is (C).** After 1 black sock is removed from the drawer, there are 9 black socks and 12 white socks. So the outcomes you want is 9. There are a total of $9 + 12 = 21$ socks in the drawer, so that is the total possible outcomes. So the probability that Sandy's socks will match is:

$$\text{Probability} = \frac{\text{Outcomes you want}}{\text{Total possible outcomes}}$$

$$\text{Probability} = \frac{9}{21}$$

$$\text{Probability} = \frac{3}{7}$$

2. **The correct answer is (B).** Originally, there are 5 cherry pops and $6 + 5 + 4 = 15$ pops total. That means that the probability of picking a cherry pop the first time is 5 out of 15. After 1 cherry pop is removed, there are 4 cherry pops left. That means that there are a total of 14 pops left. So, the probability of picking a cherry pop the second time is 4 out of 14. Multiply these 2 probabilities together to find the probability of picking a cherry pop twice in a row:

$$\text{Probability} = \frac{5}{15} \times \frac{4}{14}$$

$$\text{Probability} = \frac{1}{3} \times \frac{2}{7}$$

$$\text{Probability} = \frac{2}{21}$$

3. **The correct answer is (A).** A coin has two sides, so that is the total possible outcomes. One of them is a tail, so that is the outcome you want. The probability of flipping a tail the first time is 1 out of 2. The second time you flip the coin, there are still 2 outcomes, one of which is tails. So, the probability of flipping the coin the second time is also

1 out of 2. It is also the probability of flipping tails the third time. So the probability of flipping tails three times in a row is:

$$\text{Probability} = \frac{1}{2} \times \frac{1}{2} \times \frac{1}{2}$$

$$\text{Probability} = \frac{1}{8}$$

4. **The correct answer is (E).** After 1 blue marble is removed, there are a total of $8 + 6 + 6 = 20$ marbles in the bag. This is the total possible outcomes. You're asked for the probability of NOT removing a blue marble after this first marble is removed. Since 8 marbles are red and 6 are green, $8 + 6 = 14$ is the outcomes you want. So the probability of NOT removing a blue marble is:

$$\text{Probability} = \frac{\text{Outcomes you want}}{\text{Total possible outcomes}}$$

$$\text{Probability} = \frac{14}{20}$$

$$\text{Probability} = \frac{7}{10}$$

5. **The correct answer is (B).** There are 52 cards in a deck. After 1 is removed there are 51 cards left, so that is the total possible outcomes. There are 4 of each face value card in a deck. After a 3 is removed, there are 3 left. That is the outcomes you want. So the probability of removing a 3 is:

$$\text{Probability} = \frac{\text{Outcomes you want}}{\text{Total possible outcomes}}$$

$$\text{Probability} = \frac{3}{51}$$

$$\text{Probability} = \frac{1}{17}$$

Combinations Problems (Page 243)

1. **The correct answer is (C).** There are 6 sandwich choices, 2 side-dish choices, and 5 soft-drink choices. That means that there are 6 × 2 × 5 = 60 possible lunch specials.

2. **The correct answer is (C).** There are 4 sweater styles, 5 colors, and 4 sizes. That means that there are 4 × 5 × 4 = 80 different sweaters.

3. **The correct answer is (C).** There are 3 choices in Set A, 5 choices in Set B, and 2 choices in Set C. That means that there are 3 × 5 × 2 = 30 possible numbers.

4. **The correct answer is (E).** There are 10 pasta choices, 5 sauce choices, and 6 topping choices. That means that there are 10 × 5 × 6 = 300 basic pasta dishes.

5. **The correct answer is (D).** There are 10 car models, 6 exterior colors, and 2 interior finishes. That means that there are 10 × 6 × 2 = 120 different cars.

Permutations Problems (Page 247)

1. **The correct answer is (E).** There are 3 positions, so you need 3 slots. Since any of the 8 students can serve for one position, there are 8 possibilities for the first slot. After a student fills that position, 8 – 1 = 7 possibilities are left for the second slot. After a student fills that position, 7 – 1 = 6 possibilities are left for the last slot. That means that there are 8 × 7 × 6 = 336 possibilities.

2. **The correct answer is (B).** There are 5 seats, so set up 5 slots. Two people must sit in end seats so fill them in first: A ___ ___ ___ B. Since 2 of the 5 people are seated, you have 5 – 2 = 3 choices left for the first empty seat. After placing one of those 3 in the first empty seat, you have 3 – 1 = 2 choices for the second empty seat, and 1 for the last. This gives you a total of 3 × 2 × 1 = 6 possibilities. But that's not all. A and B could switch places like so: B ___ ___ ___ A. Each would still be in an end seat, so this fits the conditions of the question. Again there'd be 6 possibilities for seating the other 3 people. That means that there are a total of 6 + 6 = 12 possibilities.

3. **The correct answer is (D).** The code has 4 characters, so you need 4 slots. Any of the 26 letters of the alphabet can be used for the first slot. Since each number can only be used once, this means that there are $26 - 1 = 25$ choices for the second slot, $25 - 1 = 24$ for the third slot, and $24 - 1 = 23$ for the fourth slot. This means that there are $26 \times 25 \times 24 \times 23$ possible codes.

4. **The correct answer is (E).** The number is 3 digits, so you need 3 slots. You can use any of the numbers for the first digit, so you have 4 choices for the first slot. After you use one of the numbers, three are left for the second digit, and two for the third digit. So there are $4 \times 3 \times 2 = 24$ possible numbers.

5. **The correct answer is (E).** The number of possibilities for 4 people sitting in 5 seats is actually the same as the number of possibilities for 5 people sitting in 5 seats ($5 \times 4 \times 3 \times 2 \times 1 = 120$), since which seat to leave empty can be thought of as an option. If the first seat is left empty, then there are 4 possibilities for the second seat, 3 for the third, 2 for the fourth, and 1 for the fifth. This is a total of $4 \times 3 \times 2 \times 1 = 24$ possibilities. There are 24 more arrangements when the second seat is left empty, 24 when the third seat is left empty, 24 when the fourth seat is left empty, and 24 when the fifth is left empty. This is a total of $24 + 24 + 24 + 24 + 24 = 120$.

chapter 12

Go Figure

UNDERSTANDING GEOMETRY WORD PROBLEMS

Some word problems deal with shapes like rectangles, squares, or circles. To answer them you need to be able to translate the information you are given, and also know some key facts about the figures.

I'm in bad shape when it comes to these kinds of problems, so I figure I'll learn something in this chapter.

Example: The length of a rectangle is 2 less than twice its width. If the perimeter of the rectangle is 26 feet, how many feet is its length?

Solution: You're given information about the length of the rectangle as compared to the width. In other words, the length is described in terms of the width. This should send your translating sense a tingle, since you know that it's often possible to represent one value in terms of another algebraically.

Let the width be *w*. You could let it be *x*, or *n*, or any other variable you like but there's a good reason to use *w* here. Since *w* clearly stands for width, when you solve for it you're more likely to realize that you've found the width and not necessarily the whole answer to the question. In this case, you're asked for the length, so once you find *w* you'll need to plug it into the variable expression that represents length.

So what is that expression? You're told that the length is 2 less than twice the width. Since width is *w*, twice the width is $2w$. Two less than this is $2w - 2$, so this is the variable expression for the length.

You've got expressions for the length and width, so now it's time for the geometry part of the question. You're told that the perimeter of the rectangle is 26 feet. You need to know that the formula for perimeter of a rectangle is $2(l + w)$, where *l* is its length and *w* its width. If you plug your variable expressions into this formula, you have an equation with 1 variable, so you can solve for its value:

$$P = 2(l + w)$$
$$26 = 2(2w - 2 + w)$$
$$26 = 2(3w - 2)$$
$$26 = 6w - 4$$
$$30 = 6w$$
$$5 = w$$

You've found the value of *w*, which is the width of the rectangle. If this were a multiple-choice question, you could bet that 5 would be one of the wrong answers, there to tempt a careless student rushing to finish. Remember that the question asks for the length! The length is $2w - 2$, or $2(5) - 2 = 8$.

WISE POINT

There's another formula for rectangles involving *l* and *w*, the area formula: $A = lw$. It's very easy to mix up the area and perimeter formulas, so be careful. If you do all the translating work correctly but then plug your numbers into the wrong formula, you're a goner.

That's so unfair! Math can be so unforgiving.

Example: The sides of Square A are doubled to create Square B. If the area of Square A is *x* square feet, what is the area of Square B, in terms of *x*?

Solution: That's an easy one, huh? If you double the side lengths, you double the area, right? Wrong!

Think about it! Suppose that the sides of Square A were 5 feet each. The area of a square is equal to s^2, where *s* is a side of the square. This means that the area of Square A is $5^2 = 25$ square feet. The sides of Square A are doubled to make Square B, so they are $2(5) = 10$ feet. This means that the area of Square B is $10^2 = 100$ square feet. Since $4(25) = 100$, the area of Square B is four times as big as that of Square A. Since the area of Square A is *x*, the area of Square B is $4x$ square feet.

In a question like this, you could always make up numbers to see what happens. We'll give you a little math fact that can spare you this work. The fact is that the ratio of the areas of proportional figures is equal to the square of the ratios of their linear measurements. What this means in plain English is that if you increase both linear measurements (length, width, side, radius) by a factor of x, you increase the area (which is a square measurement since it's the result of multiplying two numbers) by a factor of x^2. So if you had tripled the square's sides, the resulting area would be $3^2 = 9$ times as big; if you multiplied them by 4, the area would be $4^2 = 16$ times as big; and so on.

I wish my allowance increased like that.

WISE NOTE

Another way to look at this problem is to see that since each side of the square is doubled (or multiplied by 2), its area becomes $(2 \times 5)(2 \times 5) = (2 \times 2)(5 \times 5) = 4(5 \times 5)$. In other words, you can break out the factors by which the sides are increased and multiply them to see what the overall area is multiplied by. This is helpful in cases of squares or rectangles where the sides are multiplied by different numbers.

Get Wise!

Time to shape up and work out the following geometry word problems.

1. The width of a rectangle is equal to 1 more than triple its length. If the perimeter of this rectangle is 26 feet, what is its width?

(A) 3 feet   (D) 10 feet

(B) 8 feet   (E) 30 feet

(C) 9 feet

They say that change is good, but I think I'd be better off if the figures just stayed the way they are. That would make my life a lot easier right now!

2. The length of a rectangle is halved and its width is doubled. If the area of the original rectangle is A, what is the area of the new rectangle?

(A) $\dfrac{A}{2}$ (D) $2\dfrac{1}{2}A$

(B) A (E) $4A$

(C) $2A$

3. The length of one side of a square is decreased by 2, and the length of an adjacent side is tripled. If the perimeter of the resulting rectangle is 28 feet, what is its area?

(A) 2 square feet

(B) 4 square feet

(C) 12 square feet

(D) 16 square feet

(E) 24 square feet

4. The length of a rectangle is four times its width. If the area of the rectangle is 36 square feet, how many feet is its perimeter?

(A) 3   (D) 26

(B) 12   (E) 30

(C) 15

5. The radius of a circle is tripled to create a new, larger circle. If the area of the original circle is x square feet, what is the area of the new circle?

(A) $9x$ square feet

(B) $3x^2$ square feet

(C) x^2 square feet

(D) $2x$ square feet

(E) $4x$ square feet

How Wise?

Check your answers on page 272.

TRANSLATING WORDS TO PICTURES

A lot of what we've done in this book has been translating English into algebra. Now we're going to spend some time translating from English into pictures.

That should be fun. I started to get tired of just letters and numbers. Now I can use my creativity and artistic talent.

When a geometry word problem describes a picture but doesn't give you a diagram, draw one yourself. It doesn't have to be a masterpiece. You're not going to be graded on the perfect roundness of your circles or the exact lengths of your measurements. You just want a sketch that shows you what the situation looks like and helps you organize the information you are given.

It's important to understand some key terms to translate accurately. Aside from that, the key is to draw what you're told without making any additional assumptions.

Who would do that? Don't you know that saying about what making an assumption makes out of you?

Example: A circle is inscribed within a square. If the area of the square is 64 square feet, what is the area of the circle?

Solution: You've got your pencil and a blank white piece of paper. You're ready to draw but then the word **inscribed** stops you in your tracks. You know how to draw a circle and a square, but what on earth does "inscribed" mean?

Well, "inscribed" means to draw the circle inside the square so that the two figures touch in as many places as possible. Your sketch should look something like this:

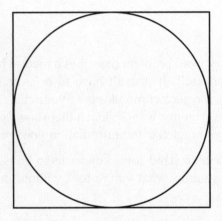

Now that you have your basic sketch, you can start filling in information. You're told that the area of the square is 64 square feet. Since the area of a square is equal to s^2, where s is a side of the square, each side of this square is 8 feet. You can mark this on the figure. As you do, notice that the diameter of the circle (the chord running through its center) is equal to a side of the square. This means that it also has a length of 8 feet.

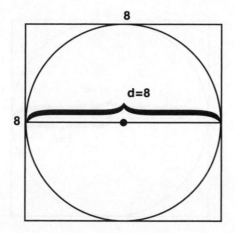

You're asked to find the area of the circle. The area of a circle is equal to πr^2, where r is a radius of the circle. The radius of a circle is equal to half its diameter. Since you know that the diameter is 8, the radius must be 4.

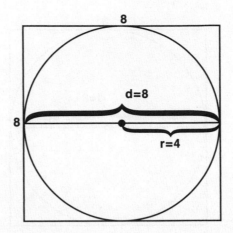

Since the radius is 4, the circle's area is $\pi(4^2) = 16\pi$.

Oh, I get it—it's just like figuring out the biggest pizza that could fit inside a square pizza box.

WISE POINTS

* *Inscribed* means to draw within, so that as many points as possible are touching.
* *Circumscribed* means to draw around, so that as many points as possible are touching.
* When you get figures with more than one shape involved, the key is to find the pieces that they have in common.

Example: Jennifer sets the margins for documents on her printer. She sets the left and right margins to 1 inch and the top and bottom margins to 1.25 inches. If she prints on standard paper that is 8.5 inches wide and 11 inches long, what is the area of the printable region of a piece of paper?

Solution: This one is a little more straightforward. The question deals with a rectangular sheet of paper, 8.5 by 11 inches. Start by sketching:

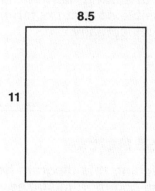

Now start filling in the information. You're told that the left and right margins are 1 inch, so mark that in. The top and bottom margins are 1.25 inches, so mark that in. Now you can sketch in the rectangular area resulting from these margins. This is the printable area.

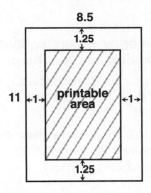

Since the printable region is a rectangle, its area will be its length times its width. You aren't given these measurements but you can use your sketch to find them. The length of the printable region is the original length minus the top and bottom margins: $11 - 2(1.25) = 11 - 2.5 = 8.5$. The width of the printable region is the original width minus the left and right margins: $8.5 - 2(1) = 8.5 - 2 = 6.5$. This makes the area of the printable region $(8.5)(6.5) = 55.25$ square inches.

I know all about setting margins. I like to make mine as big as possible so that my term papers seem longer than they actually are.

WISE NOTE

Sketching this diagram helps you avoid some common mistakes. One is to subtract the wrong margins from each measurement. The other is to subtract only one margin from each dimension.

Get Wise!

Picture this—and try the following problems that describe pictures.

1. A circle is circumscribed about a square. If the area of the circle is 81π square feet, what is the perimeter of the square?

    (A) $9\sqrt{2}$ **(D)** 81 feet

    (B) $18\sqrt{2}$ feet   **(E)** 162 feet

    (C) $36\sqrt{2}$ feet

2. The Slotnicks want to install a round swimming pool in their back-yard. Their yard is a square with a perimeter of 96 feet. There must be at least 2 feet of clearance around each edge of the pool. What is the area, in square feet, of the largest pool that could be installed?

    (A) 20π **(D)** 400π

    (B) 100π **(E)** 484π

    (C) 121π

3. Catherine has a framed photograph of her favorite movie star. The frame measures 8 inches by 10 inches around its outer perimeter. The frame is a 1.5-inch thick width of wood surrounding an insert in which the photo is displayed. What is the area, in square inches, of the wood used to make the frame?

    (A) 24.75   **(D)** 55.25

    (B) 35   **(E)** 80

    (C) 45

If Catherine really were a fan, she'd have a big poster of the star, or one of those cool life-sized cutouts.

4. A pool table is covered when not in use. The pool table is rectangular, with dimensions 4 feet by 8 feet. The cover for the pool table extends by 1 foot on each side. What is the area of the cover for the pool table in square feet?

(A) 13

(B) 28

(C) 32

(D) 45

(E) 60

5. A square is inscribed within a circle whose area is 36π square feet. What is the perimeter of the square in feet?

(A) $6\sqrt{2}$

(B) 12

(C) $24\sqrt{2}$

(D) 72

(E) 144

How Wise?

Check your answers on page 274.

A Word to the Wise

★ Translate geometry word problems carefully, representing one measurement in terms of another. Then make sure that you plug these variable expressions into the correct geometry formula.

★ The ratio of the areas of proportional figures is equal to the square of the ratio of their corresponding linear measurements.

★ When a geometry question describes a picture, make yourself a sketch. Mark it up with the given measurements and any other information that this leads to.

★ To *inscribe* means to *draw within*; to *circumscribe* means to *draw around*.

★ When you get figures with more than one shape involved, the key is to find the pieces that they have in common.

ANSWERS TO CHAPTER 12: PRACTICE EXERCISES

Geometry Word Problems (Page 261)

1. **The correct answer is (D).** You're given the width of the rectangle in terms of its length, so let the length be l. The width is 1 more than triple the length; triple the length translates to $3l$, so 1 more than this is $3l + 1$. The formula for perimeter is $P = 2(l + w)$. Plug in your algebraic expressions for the length and width. This gives you an equation with one variable, so you can solve for its value:

$$P = 2(l + w)$$
$$26 = 2(l + 3l + 1)$$
$$26 = 2(4l + 1)$$
$$26 = 8l + 2$$
$$24 = 8l$$
$$3 = l$$

Thus, the width is equal to $3l + 1$, or $3(3) + 1 = 10$ feet.

2. **The correct answer is (B).** The area of a rectangle is equal to lw, where l is the length and w the width. So area A of the original rectangle can be represented as $A = lw$. Halving its length translates to $\frac{1}{2} \times l$. Doubling its width translates to $2 \times w$. This makes the resulting area $\left(\frac{1}{2} \times l\right)(2 \times w) = \left(\frac{1}{2} \times 2\right)(l \times w) = 1(lw) = lw = A$. So, the resulting area is the same as the original area.

3. **The correct answer is (E).** A square has 4 equal sides, so each can be represented as s. Decreasing a side by 2 translates to $s - 2$. Tripling a side translates to $3s$. Since these sides are adjacent, one is the length and one is the width. The perimeter of a rectangle is $2(l + w)$, where l is length and w is width. Plug in your variable expressions.

This gives you an equation in one variable so you can solve for its value.

$$P = 2(l + s)$$
$$28 = 2\left[(s - 2) + (3s)\right]$$
$$28 = 2(4s - 2)$$
$$28 = 8s - 4$$
$$32 = 8s$$
$$4 = s$$

Since $s = 4$, the length and width of the rectangle are $s - 2 = 4 - 2 = 2$, and $3s = 3(4) = 12$. The area of a rectangle is lw, so the area of this rectangle is $(2)(12) = 24$ square feet.

4. **The correct answer is (E).** You're given the length of the rectangle in terms of its width, so let the width be w. The length is four times the width, or $4w$. The formula for area is $A = lw$. Plug in your algebraic expressions for the length and width. This gives you an equation with one variable, so you can solve for its value:

$$A = lw$$
$$36 = (4w)(w)$$
$$36 = 4w^2$$
$$9 = w^2$$
$$\sqrt{9} = \sqrt{w^2}$$
$$\pm 3 = w$$

So, w must be either 3 or –3. Since length and width measurements must be positive, $w = 3$. Thus, the width of the rectangle is $w = 3$, and the length is $l = 4w = 4 \times 3 = 12$. The perimeter is equal to $2(l + w)$, so the perimeter of this rectangle is $2(12 + 3) = 2(15) = 30$ feet.

5. **The correct answer is (A).** The area of a circle is equal to πr^2, where r is the radius of the circle. The radius is a linear measurement. That means that multiplying the radius by a factor of 3 indicates multiplying the area by a factor of $3^2 = 9$. Since the area of the original circle is x square feet, the area of the new larger circle is $9x$ square feet.

Problems That Describe Pictures (Page 269)

1. **The correct answer is (C).** Circumscribed means drawn around, so the square is inside the circle touching it at each of its four corners. Make a sketch like the one here, and then mark it up with the information you're given.

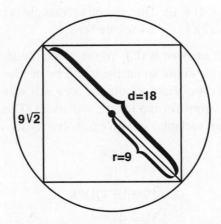

The area of a circle is πr^2, where r is the radius of the circle. Since the area of this circle is 81π, its radius is $\sqrt{81} = 9$. Note that the diameter of the circle is equal to the diagonal of the square. This means that the diagonal is $2(9) = 18$. The sides of a square times radical 2 equal its diagonal. This means that the sides of the square are $9\sqrt{2}$. The perimeter of a square is equal to $4s$, where s is a side of the square. So the perimeter of this square is $4\left(9\sqrt{2}\right) = 36\sqrt{2}$ feet.

2.   **The correct answer is (B).** The yard is square, and the Slotnicks want to put a round swimming pool in it. There has to be 2 feet between the pool and the edge of the yard. This means that your sketch should look something like the one shown here. Use it to organize your information.

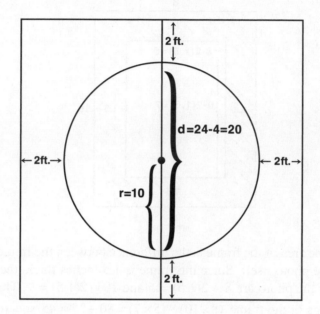

The perimeter of the square yard is 96 feet. Since the perimeter of a square is equal to 4*s*, a side of the square is equal to 96 ÷ 4 = 24 feet. Since there must be 2 feet of clearance all around, the diameter of the circle is equal to the side of the square, minus 2(2) = 4 feet: 24 − 4 = 20. This means that its radius is 10 feet. The area of a circle is equal to πr^2, so the area of the pool is $\pi(10^2) = 100\pi$ square feet.

3. **The correct answer is (C).** The framed picture has the dimensions 8 inches by 10 inches. The frame itself is 1.5 inches thick all around, so your sketch should look something like the one here. Use it to organize your information.

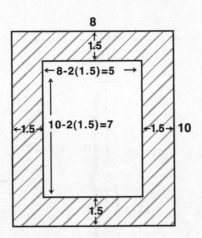

The area of the frame is the difference between the framed photo and the photo itself. Since the frame is 1.5 inches thick, the dimensions of the photo are $8 - 2(1.5) = 5$ and $10 - 2(1.5) = 7$. This makes the area of the frame $(8 \times 10) - (5 \times 7) = 80 - 35 = 45$ square inches.

4. **The correct answer is (E).** The pool table has the dimensions 4 feet by 8 feet. The cover extends past the pool table by 1 foot on each side all around, so your sketch should look something like the one here. Use it to organize your information.

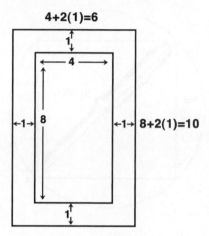

Since the cover extends beyond the pool table by 1 foot on each side, its dimensions are $4 + 2(1) = 6$ and $8 + 2(1) = 10$. This makes its area $(6)(10) = 60$ square feet.

5. **The correct answer is (C).** Inscribed means drawn within, so the square is inside the circle touching it at each of its four corners. Make a sketch like the one here, and then mark it up with the information you're given.

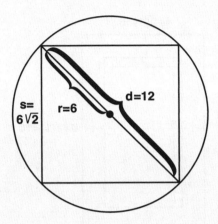

The area of the circle is 36π square feet. This means that its radius is 6 and its diameter is 12. The diameter of the circle is equal to the diagonal of the square. The sides of a square times radical 2 equal its diagonal. This means that the sides of the square are $6\sqrt{2}$. This makes its perimeter $(6\sqrt{2}) \times 4 = 24\sqrt{2}$.

chapter 13

Odds and Ends

You're almost there! This is the last chapter, and it will focus on word problems that come up a lot but don't really fit in any of the other chapters of the book.

This chapter sounds like the island of misfit toys. Do you think that no math problem is truly happy until it finds a child to solve it?

SIMPLE INTEREST PROBLEMS

As you probably know, interest is a charge for borrowed money or the money earned from an investment. Most word problems stick with the rosier side of things and focus on interest earned through investments.

In simple interest problems, the interest is always calculated on the original investment. In other words, you don't add this year's interest to next year's principal and calculate the interest on that new total. The bad news is that this isn't much like real life, where compound interest is the norm. The good news is that it's quicker and easier to calculate simple interest. There's a four-piece formula that you use to calculate simple interest:

Interest = Principal × Rate × Time

or

$I = prt$

As long as you have three of the pieces you can always solve for the fourth.

Example: Joe invested $5,000 into a mutual fund yielding 6% interest annually. If he left the money in for 3 years, how much interest did he earn?

Solution: Use the four-piece formula. Identify the given information to see where it fits into the formula. The principal is the original amount put into the investment. In this case, it's $5,000. The rate is the percent at which the interest is paid. In this case, it's 6%. Since you'll be doing calculations, it makes sense to write the percent as a decimal, in this case .06. The time is how long the money is invested for. Here, it's 3 years. Plug these values into the formula to solve for the amount of interest earned:

$$I = prt$$
$$I = (\$5,000)(.06)(3)$$
$$I = \$900$$

WISE NOTE

This question specifies that the interest is paid annually, which means each year. Most questions involve annual interest. This means that if the money is invested for a time period that's not an increment of a year, you'll need to convert it. In other words, if you were told that the money was invested

for 6 months, you'd convert this to $\frac{1}{2}$ or .5

years.

Be sure to read the question carefully. This question could have asked for the value of the investment after 3 years, instead of the interest earned in 3 years. This would be the amount of the original investment, or the principal, plus the interest it earned over the 3 years. You'd solve for the interest as above, and then add it to the principal: $5,000 + $900 = $5,900

You earn $900 and don't have to work for it? I like the sound of that!

Some interest questions are a little more complicated. They involve multiple investments and require you to set up and solve an equation.

Multiple investments—do I need to call my broker for this?

Example: The Steddins invest $10,000 into an account that earns 4% annually. How much additional principal must they invest into an account earning 10% annually so that the total annual income earned in one year is 6% of the total investment?

Solution: First, figure out just what the question is looking for. The amount of interest earned from the 4% investment, plus the amount of interest earned from the 10% investment, must equal the interest that would be earned by placing the total amount of principal into an account earning 6%.

Now set up an equation to reflect this situation. You're looking for the amount that must be invested at 10%, so let it be x. Use the formula $I = prt$ to figure out the interest earned from each investment.

$$\text{Int. 4\% acct.} + \text{Int. 10\% acct.} = \text{6\% int. total investment}$$
$$(\$10,000)(.04)(1) + (x)(.10)(1) = (\$10,000 + x)(.06)(1)$$
$$\$400 + .10x = \$600 + .06x$$
$$.10x = \$200 + .06x$$
$$.04x = \$200$$
$$x = \$5,000$$

It sounds like it's a lot of work to make money, but once you have it, you can sit back and make even more money—by getting interest. Now that's more like it!

Get Wise!

Ready to try some simple interest problems? These are a breeze!

1. Amy deposited $3,000 into an investment for 4 years. If she earned a total of $360 interest during this period, what was the annual interest rate?

   (A) 3%   **(D)** 9%

   (B) 4%   **(E)** 12%

   (C) 6%

2. A certain investment pays an interest rate of 8% annually. If an investor wanted to earn interest of $1,600 each year, how much more principal would need to be invested?

   (A) $128   **(D)** $20,000

   (B) $200   **(E)** $200,000

   (C) $2,000

3. Ms. Kleinman plans to invest a total of $25,000 into two different mutual funds. One account pays 5% interest annually, and the other pays 11% annually. Her financial planner advises her that for tax purposes she shouldn't earn more than $2,150 in interest this year. How much should Ms. Kleinman invest in each account?

   (A) 5% fund: $8,000; 11% fund: $17,000

   (B) 5% fund: $10,000; 11% fund: $15,000

   (C) 5% fund: $12,500; 11% fund: $12,500

   (D) 5% fund: $15,000; 11% fund: $10,000

   (E) 5% fund: $17,000; 11% fund: $8,000

4. After 5 years, $8,000 invested in an interest-bearing account is worth a total of $10,400. What was the annual interest rate for this account?

   (A) 3%   **(D)** 26%

   (B) 6%   **(E)** 30%

   (C) 13%

5. Mr. Big invests $5,000 into an account that earns 5% annually. How much additional principal must he invest into an account earning 10% annually so that the total annual income earned in one year is 7.5% of the total investment?

   (A) $2,500
   (B) $3,000
   (C) $5,000
   (D) $8,000
   (E) $10,000

How Wise?

Check your answers on page 293.

COIN AND BILL PROBLEMS

More money problems ahead! In coin and bill problems, you're given information about the relative numbers of each denomination, as well as their total value. Your job is to figure out how many of each coin or bill there is.

I can relate to money problems—have you seen the size of my allowance? My main problem is that I don't have nearly enough of either coins OR bills.

Example: Zach looks under the sofa cushions and finds a total of $2.00 in nickels and dimes. If Zach found twice as many dimes as nickels, how many dimes did he find?

I feel your pain, Zach. I'm not ashamed to admit that I've frisked the sofa a time or two in search of some extra cash. By the way, the seats in the car are another good place to look.

Solution: The key to these questions is to account for the value of each coin or bill. It's also important to use the same units (dollars or cents) throughout.

Zach found twice as many dimes as nickels. If you let the number of nickels he found be x, that means that the number of dimes is twice that, or $2x$. After you've found a way to represent the number of each coin with a variable, you can write an equation.

You know that the total value of the nickels and dimes is $2.00. To find the value of each type of coin, you multiply the number of the coin by its value. A nickel is worth 5 cents, so all the nickels are worth $5x$. A dime is worth 10 cents, so all the dimes are worth $10(2x) = 20x$. If you add the value of the nickels to the value of the dimes, you can set them equal to $2.00. Since you expressed the values of the coins in cents, you should express the value of the total in cents as well. Since a dollar is worth 100 cents, $2.00 is equal to 200 cents.

$$5x + 20x = 200$$
$$25x = 200$$
$$x = 8$$

Since $x = 8$, there are 8 nickels and $2(8) = 16$ dimes. You can check your work by multiplying through and seeing if it adds up.

$$8(5) + 16(10) = 200$$
$$40 + 160 = 200$$
$$200 = 200$$

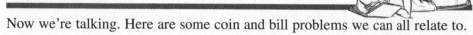

Get Wise!

Now we're talking. Here are some coin and bill problems we can all relate to.

1. John has a total of $300 in his wallet, in 10- and 20-dollar bills. If John has three times as many tens as twenties, how much money does he have in tens?

(A) $40   (D) $180

(B) $60   (E) $360

(C) $120

2. Becca has a total of $3.95 in nickels, dimes, and quarters. If she has twice as many nickels as dimes, and 5 fewer dimes than quarters, how many quarters does Becca have?

(A) 6   (D) 29

(B) 11   (E) 60

(C) 12

3. A jar contains half as many nickels as dimes, and twice as many quarters as dimes. If the jar contains a total of $11.25, how much money does it contain in dimes?

(A) $0.45   (D) $1.80

(B) $0.90   (E) $9.00

(C) $1.20

Whatever the answer, that's not much of a change jar. I knew this guy who put all his change in a jar, and when it was full he had enough to buy a TV.

4. When Arlene closes out her cash register at the end of her shift, it contains a total of $645 in cash. All of the bills are fives, tens, or twenties. If there are twice as many twenties as fives, and 7 fewer tens than twenties, how many tens does it contain?

(A)	4	**(D)**	22
(B)	11	**(E)**	29
(C)	15		

5. A purse contains $2.25 in nickels and quarters. If there are one-fourth as many quarters as nickels, how much money does it contain in quarters?

(A)	$0.50	**(D)**	$1.25
(B)	$0.75	**(E)**	$1.75
(C)	$1.00		

How Wise?

Check your answers on page 295.

AGE PROBLEMS

When you're young you want to be or at least look older, but my mom tells me that one day the opposite will be true. Guess that's what you mean by age problems?

Sort of, Chi. Age problems talk about the ages of one or more people. Their present ages are compared to their past or future ages. We'll show you how to handle them without it seeming like it's taking years off your life.

Example: Jane is 10 years older than Lydia. In 5 years, Jane will be twice as old as Lydia. How old is Jane now?

Solution: Read the information carefully and translate it into an equation. Be especially careful when translating information about past or future ages.

You're told that Jane is 10 years older than Lydia. If you let Lydia's current age be x, then Jane's current age is $x + 10$.

In 5 years Jane will be twice as old as Lydia. Be careful here. To find their ages in 5 years you need to add 5 years to each of their current ages. So Lydia's age in 5 years is $x + 5$, and Jane's age in 5 years is $x + 10 + 5 = x + 15$.

Since Jane will be twice as old as Lydia in 5 years, Jane's age 5 years from now is equal to twice Lydia's age 5 years from now. Write this as an equation and solve for x:

$$\text{Jane's age in 5 years} = 2(\text{Lydia's age in 5 years})$$
$$x + 15 = 2(x + 5)$$
$$x + 15 = 2x + 10$$
$$15 = x + 10$$
$$5 = x$$

So you've found the value of x, which means that Lydia's current age is 5. Remember though that the question asks for Jane's current age. This is represented by the expression $x + 10$, so Jane's current age is $5 + 10 = 15$.

WISE NOTE

Remember, solving for the variable doesn't always give you the answer the question asks for. You might need to use the value of the variable to find another value. Always reread the question to make sure that you're answering the question that's asked.

Get Wise!

Brace yourself, these age problems make up the last set of practice questions!

1. Fric is four times as old as Frac. Four years ago, Fric was ten times as old as Frac. What is Fric's current age?

   (A) 2   (D) 20

   (B) 6   (E) 24

   (C) 12

2. Tina is 11 years younger than Telly. If the sum of their ages is 19, how old is Telly?

   (A) 4   (D) 13

   (B) 8   (E) 15

   (C) 11

3. Meryl is 6 years younger than Mia. Eleven years ago, Mia was three times as old as Meryl. How old is Mia now?

   (A) 9   (D) 20

   (B) 11   (E) 28

   (C) 14

4. Damon is one-fourth as old as his brother David. In 2 years, David will be three times as old as Damon. How old is David?

   (A) 4   (D) 18

   (B) 6   (E) 20

   (C) 16

5. Mike is 6 years younger than Al. In 2 years the sum of their ages will be 20. How old is Al now?

   (A) 5   (D) 14

   (B) 11   (E) 16

   (C) 12

How Wise?

Check your answers on page 298.

A Word to the Wise

- ★ To figure simple interest, use the formula: Interest = Principal × Rate × Time, or $I = prt$.

- ★ Some interest questions involve multiple investments and require you to set up and solve an equation, using the interest formula.

- ★ The key to coin and bill problems is to account for the value of each coin or bill. It's also important to use the same units (dollars or cents) throughout.

- ★ To answer an age problem, read the information carefully and translate it into an equation. Be especially careful when translating information about past or future ages.

ANSWERS TO CHAPTER 13: PRACTICE EXERCISES

Simple Interest Problems (Page 283)

1. **The correct answer is (A).** Use the four-piece formula. The principal is $3,000, the time is 4 years, and the interest is $360. Plug these values into the formula to solve for the rate.

$$I = prt$$
$$\$360 = (\$3,000)(r)(4)$$
$$\$360 = \$12,000r$$
$$\frac{\$360}{\$12,000} = \frac{\$12,000r}{\$12,000}$$
$$.03 = r$$
$$3\% = r$$

2. **The correct answer is (D).** Use the four-piece formula. The rate is 8%, the time is 1 year, and the interest is $1,600. Plug these values into the formula to solve for the principal.

$$I = prt$$
$$\$1,600 = (p)(.08)(1)$$
$$\$1,600 = .08p$$
$$\frac{\$1,600}{.08} = \frac{.08p}{.08}$$
$$\$20,000 = p$$

3.   **The correct answer is (B).** Let the amount invested at 5% be x. Since the total amount of the investment was $25,000, the amount invested at 11% must be $25,000 – x. The total income from these investments should be $2,150. Use the simple-interest formula to write an expression for the interest earned in each account, and then sum them to $2,150.

$$(x)(.05)(1) + (\$25,000 - x)(.11)(1) = \$2,150$$
$$.05x + \$2,750 - .11x = \$2,150$$
$$-.06x + \$2,750 = \$2,150$$
$$-.06x = -\$600$$
$$\frac{-.06x}{-.06x} = \frac{-\$600}{-.06}$$
$$x = \$10,000$$

Since $10,000 is invested at 5%, the remaining $25,000 – $10,000 = $15,000 must be invested at 11%.

4.   **The correct answer is (B).** Use the four-piece formula. The principal is $8,000 and the time is 5 years. Since the original investment of $8,000 is now worth $10,400, the interest earned is $10,400 – $8,000 = $2,400. Plug these values into the formula to solve for the rate.

$$I = prt$$
$$\$2,400 = (\$8,000)(r)(5)$$
$$\$2,400 = \$40,000r$$
$$\frac{\$2,400}{\$40,000} = \frac{\$40,000r}{\$40,000}$$
$$.06 = r$$
$$6\% = r$$

5. **The correct answer is (C).** The amount of interest earned from the 5% investment, plus the amount of interest earned from the 10% investment must equal the interest that would be earned by placing the total amount of principal into an account earning 7.5%. Set up an equation to reflect this situation. You're looking for the amount that must be invested at 10%, so let it be x. Use the formula $I = prt$ to figure out the interest earned from each investment.

$$(\$5,000)(.05)(1) + (x)(.10)(1) = (\$5,000 + x)(.075)(1)$$
$$\$250 + .10x = \$375 + .075x$$
$$.10x = \$125 + .075x$$
$$.025x = \$125$$
$$x = \$5,000$$

Coin and Bill Problems (Page 287)

1. **The correct answer is (D).** Let the number of twenties that John has be x. This means that he has $3x$ tens. Set up an equation summing the total value of cash for each type of bill to 300.

$$10(3x) + 20(x) = 300$$
$$30x + 20x = 300$$
$$50x = 300$$
$$x = 6$$

Since John has $3x$ tens, he has $3(6) = 18$ tens, for a total of $18(\$10) = \$180$.

2.   **The correct answer is (B).** Let the number of dimes that Becca has be x. This means that she has $2x$ nickels. Since she has 5 fewer dimes than quarters, she has $x + 5$ quarters. Set up an equation summing the total value of cash for each type of coin to 395.

$$5(2x) + 10(x) + 25(x + 5) = 395$$
$$10x + 10x + 25x + 125 = 395$$
$$45x + 125 = 395$$
$$45x = 270$$
$$x = 6$$

Since Becca has $x + 5$ quarters, she has $6 + 5 = 11$ quarters.

3.   **The correct answer is (D).** Let the number of nickels be x. This means that the number of dimes is $2x$. Since there are twice as many quarters as dimes, there are $4x$ quarters. Set up an equation summing the total value of cash for each type of coin to $\$11.25 = 1,125$.

$$5(x) + 10(2x) + 25(4x) = 1,125$$
$$5x + 20x + 100x = 1,125$$
$$125x = 1,125$$
$$x = 9$$

Since there are $2x$ dimes, there are $2(9) = 18$ dimes, for a total of $18(\$0.10) = \$1.80$.

4. **The correct answer is (C).** Let the number of fives be x. This means that the number of twenties is $2x$. Since there are 7 fewer tens than twenties, there are $2x - 7$ tens. Set up an equation summing the total value of cash for each type of bill to 645.

$$5(x) + 10(2x - 7) + 20(2x) = 645$$
$$5x + 20x - 70 + 40x = 645$$
$$65x - 70 = 645$$
$$65x = 715$$
$$x = 11$$

Since there are $2x - 7$ tens, there are $2(11) - 7 = 22 - 7 = 15$ tens.

5. **The correct answer is (D).** Let the number of quarters be x. That means that the number of nickels is $4x$. Set up an equation summing the total value of cash for each type of coin to 225.

$$5(4x) + 25(x) = 225$$
$$20x + 25x = 225$$
$$45x = 225$$
$$x = 5$$

Since there are 5 quarters, $5(\$0.25) = \$1.25$ of the money is in quarters.

Age Problems (Page 291)

1. **The correct answer is (E).** Let Frac's age be x. That means that Fric's age is $4x$. Four years ago Fric was ten times as old as Frac. This means that ten times Frac's age 4 years ago is equal to Fric's age 4 years ago. Set up an equation to show this and solve for x.

$$10(x-4) = 4x - 4$$
$$10x - 40 = 4x - 4$$
$$10x = 4x + 36$$
$$6x = 36$$
$$x = 6$$

Fric's current age is $4x$, or $4(6) = 24$.

2. **The correct answer is (E).** Let Tina's age be x. Since she's 11 years younger than Telly, his age is $x + 11$. Their ages sum to 19, so set up an equation and solve for x.

$$x + x + 11 = 19$$
$$2x + 11 = 19$$
$$2x = 8$$
$$x = 4$$

Telly's age is $x + 11$, or $4 + 11 = 15$.

3. **The correct answer is (D).** Let Meryl's age be x. She is 6 years younger than Mia, so Mia's age is $x + 6$. Eleven years ago Mia was three times as old as Meryl. This means that three times Meryl's age 11 years ago is equal to Mia's age 11 years ago. Set up an equation to show this and solve for x.

$$3(x-11) = x+6-11$$
$$3x-33 = x-5$$
$$3x = x+28$$
$$2x = 28$$
$$x = 14$$

Mia's current age is $x + 6$, or $14 + 6 = 20$.

4. **The correct answer is (C).** Let Damon's age be x. Since it is one-fourth of David's age, David's age is $4x$. In 2 years, David will be three times as old as Damon. This means that three times Damon's age in 2 years is equal to David's age in 2 years. Set up an equation to show this and solve for x.

$$3(x+2) = 4x+2$$
$$3x+6 = 4x+2$$
$$3x+4 = 4x$$
$$4 = x$$

David's age is $4x$, or $4(4) = 16$.

5. **The correct answer is (B).** Let Mike's age be x. Since he is 6 years younger than Al, Al's age is $x + 6$. In 2 years, the sum of their ages will be 20. This means that Mike's age in 2 years plus Al's age in 2 years sums to 20. Set up an equation to show this and solve for x.

$$(x+2)+(x+6+2) = 20$$
$$x+2+x+8 = 20$$
$$2x+10 = 20$$
$$2x = 10$$
$$x = 5$$

Al's age is $x + 6$, or $5 + 6 = 11$.

Notes

Notes

Notes

# Notes